FINAL TECHNICAL REPORT

Project Period: September 1, 2006 to September 30, 2010

Advanced Wear-resistant Nanocomposites for Increased Energy Efficiency
(CPS agreement no.15015)

B. A. Cook, J. L. Harringa, and A. M. Russell
Ames Laboratory
Division of Materials Science & Engineering
Iowa State University, Ames, IA 50011-3020
(515) 294-9673
cook@ameslab.gov
harringa@ameslab.gov
russell@iastate.edu

Project partners:

Oak Ridge National Laboratory
Carpenter Powder Products, Inc.
The University of Alberta

February 2012

Acknowledgement, Disclaimer, and Proprietary Data Notice

Acknowledgement

This report is based upon work supported by the U.S. Department of Energy, Energy Efficiency and Renewable Energy, Industrial Technologies Program, Materials for Energy Efficient Industrial Processing under CPS agreement number 15015.

Research at the Oak Ridge National Laboratory was sponsored by the U.S. Department of Energy, Office of Energy Efficiency and Renewable Energy, Industrial Technologies Program, under contract DE-AC05-00OR22725 with UT-Battelle, LLC.

Research at The Ames Laboratory was sponsored by the U.S. Department of Energy, Office of Energy Efficiency and Renewable Energy, Industrial Technologies Program, under contract DE-AC02-07CH11358 with Ames Laboratory, Iowa State University.

Disclaimer

Table of Contents

Abbreviations and Acronyms 4

1.0   Executive Summary 5

   1.1   Purpose 5

   1.2 Scope 5

   1.3   Results 6

   1.4 Conclusions 8

   1.5   Recommendations 8

   1.6   Commercialization 8

2.0   Background 9

   Specific Area Being Addressed 9

   Technical Approach and Hypothesis 9

   Key project participants 10

3.0   Results 11

   Task 1: Processing Studies 11

      1.1: Optimization of TiB_2 - $AlMgB_{14}$ Ratio 11

      1.2: Microstructure 21

      1.3: 2nd and 3rd Phase Additions 33

      1.3.2: SiC Whisker Reinforcement 40

      1.3.3: Co-Mn Binder 47

      1.4: Oxygen Reduction 64

   Task 2: Development of Scale-up Technology 73

      2.1: Powder Processing 73

      2.2. Consolidation 88

   Task 3. Application-specific Testing 108

   Task 4. Commercialization 118

   Task 5: Computational Modeling 118

      5.1: Wear Modeling 118

      5.2: Electronic Structure Modeling 123

   Task 6. High Load Indentation Testing 124

4.0   Discussion 128

5.0   Accomplishments 129

6.0   Conclusions 130

7.0   Recommendations 132

8.0   References 133

   Task 1: Processing & Characterization References 133

   Task 2: Development of Scale-up Technology References 134

   Task 6: High Load Indentation Testing References 135

Abbreviations and Acronyms

AFM	Atomic Force Microscopy
ASTM	American Society for Testing Materials
at.	Atomic
BAM	$AlMgB_{14}$-based material
BAMC	$AlMgB_{14}$-TiB_2+C coating
Co-Mn	Co-16%Mn (wt. %)
DOE	Department of Energy
EDS	Energy-Dispersive Spectroscopy
h-BN	(hexagonal) boron nitride
HIP	Hot Isostatically Pressed
HP	Hot uniaxially Pressed
HT	High Temperature
Hz	Hertz
MA	mechanical alloying (solid state synthesis)
NDA	Non-Disclosure Agreement
ORNL	Oak Ridge National Laboratory
PLD	Pulsed Laser Deposition
Psi	Pounds per Square Inch
REV	Revolution(s)
S-	Laboratory-scale (high-energy) vibratory mill
SEM	Scanning electron microscopy
SPS	Spark Plasma Sintering
TB00	$AlMgB_{14}$ baseline (no TiB_2 addition)
TB50	$AlMgB_{14}$ + 50 vol% TiB_2
TB60	$AlMgB_{14}$ + 60 vol% TiB_2
TB100	100 % TiB_2
TEM	Transmission electron microscopy
T_m	melting temperature
wt.	Weight
XPS	X-ray photo spectroscopy
XRD	X-ray diffraction
Z-	Large-scale attritor mill

1.0   Executive Summary

This report summarizes the work performed by an Ames-led project team under a 4-year DOE-ITP sponsored project titled, "Advanced Wear-resistant Nanocomposites for Increased Energy Efficiency." The Report serves as the project deliverable for the CPS agreement number 15015.

1.1   Purpose

The purpose of this project was to develop and commercialize a family of lightweight, bulk composite materials that are highly resistant to degradation by erosive and abrasive wear. These materials, based on $AlMgB_{14}$, are projected to save over 30 TBtu of energy per year when fully implemented in industrial applications, with the associated environmental benefits of eliminating the burning of 1.5 M tons/yr of coal and averting the release of 4.2 M tons/yr of CO_2 into the air.   This program targeted applications in the mining, drilling, machining, and dry erosion applications as key platforms for initial commercialization, which includes some of the most severe wear conditions in industry. Production-scale manufacturing of this technology has begun through a start-up company, NewTech Ceramics (NTC). This project included providing technical support to NTC in order to facilitate cost-effective mass production of the wear-resistant boride components.

Resolution of issues related to processing scale-up, reduction in energy intensity during processing, and improving the quality and performance of the composites, without adding to the cost of processing were among the primary technical focus areas of this program. Compositional refinements were also investigated in order to achieve the maximum wear resistance. In addition, synthesis of large-scale, single-phase $AlMgB_{14}$ powder was conducted for use as PVD sputtering targets for nanocoating applications.

1.2 Scope

To accomplish the above objective, the project was divided into 5 tasks with the following specific objective and work scope:

Task 1. Processing Studies
This task involved issues related to the nature of precursor materials, powder production, consolidation, and high-temperature sintering as applied to laboratory-scale samples in the 10 to 100 gram range. In particular, we examined processing variables to enhance the properties of wear-resistant boride nanocomposites or reduce processing costs. Included in this task was a determination of the optimum ratio of phases within the $AlMgB_{14}$-TiB_2 system leading to maximum wear resistance, and an examination of various oxygen getters and ductile binder phase additions to $AlMgB_{14}$-TiB_2 for the purpose of enhancing sinterability and wear resistance. Methods were developed to decrease the energy intensity of processing, by identifying alternatives to conventional dry-milling of powders. This task also included characterization of the properties of consolidated samples, specifically characterization of microstructure, hardness, toughness, erosive and abrasive wear, and chemical reactivity.

Task 2. Development of Scale-up Technology

This task addressed the challenge of incorporating the processing refinements developed in Task 1 to large-scale production of kilogram quantities of $AlMgB_{14}$-TiB_2 powder and consolidated materials. Industrial partner Carpenter Powder Products is providing support with advanced consolidation techniques, such as their Dynaforge process, to develop improved, near-net shape components for wear testing. National Laboratory partner Oak Ridge provided support for the synthesis of powder using microwave and IR processing, and recently initiated advanced consolidation studies using spark plasma sintering (SPS).

Task 3. Application-specific testing

This task provided for evaluation of wear-resistant $AlMgB_{14}$-TiB_2 samples and components in real-world environments to determine lifetime and reliability of the component compared with current-generation materials. For example, Missouri University of Science & Technology provided field testing support for abrasive waterjet nozzle applications using their ultra-high pressure waterjet laboratory. These tests helped to define the economic and energy-saving benefits of the boride material. Greenleaf Corporation provided technical support for development of advanced bulk ceramic materials for high-speed machining applications, and agreed to provide field-testing of toughened $AlMgB_{14}$-based inserts. Testing was also initiated for severe service applications involving dry erosive wear such as sandblasting nozzles, for which these boride composites are particularly well-suited.

Task 4. Commercialization

This task provided support for efforts to commercialize the wear-resistant boride technology. Many materials-related start-up companies fail within the first 2 years of operation as a result of insufficient resources to sustain the processing developments needed to optimize the material in commercial quantities. During the nascent stage of its commercialization efforts, we provided support to NewTech Ceramics in the form of analysis and characterization of their production material, and by providing recommendations for improvements in their processing. Carpenter's consolidation technology was useful in transferring net-shape processing methods to commercial production. In addition, we actively identified and pursued new applications where substantial energy saving benefits from incorporation of this technology may be realized.

Task 5. Computational modeling

The University of Alberta led the effort to develop a robust, computational model for wear resistance in complex, multiphase ceramic composites. Parameters of the model include porosity, size and distribution of the reinforcement phase, and effect of secondary or tertiary reinforcement phases. The objective was to improve predictability of microstructure and composition for maximum resistance to wear, and to provide a fundamental basis for the observed behavior.

1.3 Results

Selected key results from this project are listed below.

- Developed a novel synthesis route to prepare ultra-hard $AlMgB_{14}$ at atmospheric pressure conditions

- Achieved a record erosion resistance of 0.14 mm³/Kg in an $AlMgB_{14}$ - $(Ti,Zr)B_2$ composite. This compares with 0.50 mm³/Kg, established as a typical erosion resistance for production-quality $AlMgB_{14}$ - TiB_2 composites.

- Discovered that addition of a third binary boride can reduce steady state erosive wear resistance to as low as 0.14 mm³/kg. This represents a factor of 5 reduction in wear rate compared with the initial objectives at the inception of this project.

- Established feasibility of pressing and sintering as a cost-effective method of producing commercial-scale boride powder.

- Developed a wet-milling process and subsequent densification technology for $AlMgB_{14}$-TiB_2 powder at Carpenter Powder Products, Inc. by the Dynaforge technique. Samples prepared near the end of the project achieved excellent uniformity, full density, and high toughness.

- Addition of 1% stearic acid was found to decrease milling time of the boride powders from 12 hours down to 2.

- Successfully prepared $AlMgB_{14}$ in large quantities (e.g., 50 cm³) by radiant heating in a graphite resistance furnace.

- Developed a computational model to investigate the influence of porosity on the solid-particle erosion of $AlMgB_{14}$-TiB_2 ceramic matrix composites.

- Developed a procedure to achieve a significant improvement in distribution of the highly ductile Co_Mn binder within the mixed-phase $AlMgB_{14}$-TiB_2 composite powder

- Validated processing of a 50 cm³ compact of $AlMgB_{14}$+60 v% TiB_2 by spark plasma sintering (SPS).

- Established that $AlMgB_{14}$/TiB_2 composites are suitable for use in acidic environments such as H_2SO_4.

The project's initial milestone objective was to achieve synthesis of an $AlMgB_{14}$-TiB_2 nanocomposite of at least 50 cm³ in total volume while meeting the following specifications: (*values in parentheses indicate typical values at the beginning of this project*):
- Total porosity < 1% (2 – 4%)
- $MgAl_2O_4$ < 4% (5 – 10%)
- FeB < 4% (10 – 15%)
- Wear rate < 0.75 mm³/kg when subjected to steady-state erosion at normal incidence by 200 μm Al_2O_3 at 77 m/s with a flow rate of 4.5 g/min at a standoff distance of 1 cm. (3.0 mm³/kg)

All of these milestones were achieved and process reproducibility was established. An updated set of milestones was subsequently established during the last year of this project, defined by a steady state wear rate of less than 0.18 mm³/kg under identical conditions for an equivalent volume of material. The project also achieved a record wear resistance value of 0.14 mm³/kg for laboratory-scale composites in dry erosion tests, which compares with 0.5 mm^3/kg for monolithic TiB_2, 0.2 mm^3/kg for RocTec 500, and 0.13 mm^3/kg for Borazon, a trade name of one of the most wear-resistant grades of cubic boron nitride.

1.4 Conclusions

This project demonstrated that the $AlMgB_{14}$–based nanocomposites offer wear resistance in bulk materials comparable to or better than state-of-the-art materials such as RocTec and Borozon (cubic BN), in addition to providing an exceptionally low friction coefficient as a result of the formation of a surface layer of boric acid. A key accomplishment of this project was to develop processing technology to produce large-scale quantities of boride nanocomposite material with properties comparable to that observed in earlier laboratory-scale samples.

Implementation of this material in key applications will result in efficiency savings in both erosion-intensive and high-performance tooling systems. It is expected that the use of this technology in other wear-intensive markets will also yield savings. These materials are projected to save over 30 TBtu of energy per year when fully implemented in industrial applications, with the associated environmental benefits of eliminating the burning of 1.5 M tons/yr of coal and averting the release of 4.2 M tons/yr of CO_2 into the air.

1.5 Recommendations

There are a few unresolved questions regarding the composite and how much further its performance could be enhanced. Current knowledge of the material's electronic structure is somewhat established, though future changes in composition could yield nanocomposites with even greater wear or frictional performance. Further R&D work would help define how far this particular material system can be advanced in terms of friction, wear, and durability.

1.6 Commercialization

The reader is referred to the section on Task 4 for specific details regarding the commercialization pathways for these materials.

2.0   Background

Specific Area Being Addressed

$AlMgB_{14}$ is a lightweight, low-friction, wear resistant material that can be used in both the large-scale bulk form and as a protective coating. This material is the only available product that combines high wear resistance with a regenerating lubricant that delivers an extremely low coefficient of friction. Initially developed by Ames Laboratory, the material is synthesized through conventional powder processing techniques. Department of Energy funding was used to support further development of the technology through materials research, tribological modeling and simulation, and end-application performance testing to facilitate the commercial implementation and thereby realize energy savings in the industry.

Technical Approach and Hypothesis

The proposed research, development, and commercialization plan was largely focused on improving materials' resistance to degradation in a variety of industrial applications. These applications were well suited for fundamental technology development and eventual technology transfer because they are inherently ubiquitous in their scope, application space, and function. Furthermore, the two applications identified offer tremendous potential for broad industrial adoption and related energy savings.

Industrial process efficiency is directly linked to the wear and degradation of materials used in these applications. The preferred route to minimizing wear and enhancing product performance lies in the application of sufficiently hard materials such that the surfaces experience lower friction and resist wear (thus prolonging component life). With reduced friction between contacting surfaces, less energy is required to combat frictional losses during start-up and/or operation of the larger system(s), thereby yielding improved energy efficiency. While the increase in efficiency per component (e.g., hydraulic pump, seal, shaft, bearings) may be modest, the large number of pumps in operation results in a substantial cumulative energy savings. Clearly, hard materials are needed, and interest is especially high for materials exhibiting so-called "superhardness," or hardness exceeding 40 GPa. Diamond, diamond-like carbon (DLC), and cubic BN all fall within this category. While each such material possesses high hardness, each also falls short in critical performance areas. For example, diamond and DLC coatings are prone to oxidation at temperatures exceeding ~ 600°C, and they can undergo undesirable chemical reactions upon contact with ferrous metals. CBN, which is relatively inert with respect to ferrous metals, is difficult to deposit by PVD at thicknesses exceeding 300 nm because of residual stresses that build up in the films.

The family of $AlMgB_{14}$-based nanocomposites emerged prior to the onset of this project and offered some potentially substantial advantages. In this particular subset of boride composites, the hardness is derived from microstructural engineering of the constituent phases, referred to as extrinsic hardness. In the sections to follow, we will identify how the concept of extrinsic hardness was developed to advance the state-of-the-art of wear-resistant bulk materials.

Key project participants

Ames Laboratory:
Bruce Cook, Alan Russell, and Joel Harringa: Conducted research that led to original discovery and corresponding patents; assisted in finding partners interested in licensing the technology; Cook and Harringa collaborated on pulsed laser deposition (PLD) coatings process development, and Russell determined PLD parameters for later coatings trials. All three developed the proper composition of the materials to increase wear-resistance. Ames Laboratory provided some of the material used as targets for physical vapor deposition, along with analysis of coatings before and after dynamometer testing, and ongoing technical support.

Oak Ridge National Laboratory:
Steve Nunn: provided support in the areas of large-scale powder processing, and reaction sintering of the constituent materials.

Greenleaf Corporation:
Jason Goldsmith: Provided standard cutting tool inserts for deposition substrates and in-house and external cutting tests utilizing coated tools. Goldsmith developed appropriate testing protocols for various machining scenarios and provided interpretation of results. He also developed energy-saving models to predict the impact in high-speed machining applications and investigated additional applications for the technology.

University of Alberta:
Dongyang Li: provided for development of a robust mathematical model of the wear resistance of ceramic-reinforced ceramic matrix composites, specifically designed to predict optimal range of wear resistance in complex $AlMgB_{14}$-based materials. The model will take into account volume fraction and size distribution of reinforcement and effects of porosity.

Carpenter Powder Products, Inc.:
David Novoknak: CPP was a cost-sharing industrial partner and provided cost-effective consolidation solutions for producing fully dense, net shape samples using their Dynaforge process, which as of October, 2010, was undergoing a major upgrade. Greenleaf Corporation, a global supplier of high-performance cutting tool solutions, provided technical support and evaluation of advanced boride composites developed by this program.

3.0   Results

The following provides a description of the research and development efforts related to the processing and characterization of wear-resistant $AlMgB_{14}$-TiB_2 composites, as related to commercialization and scale-up of the technology. Technical issues such as the nature of starting materials, powder production, consolidation, high temperature sintering, and the properties of the consolidated samples are presented to describe the production of high wear resistant materials and to determine the most cost-effective and energy-efficient processing routes for this family of composites.

Task 1: Processing Studies

Overview of Section 1

The hardness and wear resistance of the $AlMgB_{14}$-TiB_2 composites vary with the concentration of TiB_2. Previous studies showed greater than rule-of-mixtures hardness and toughness for the composites, and the project's search for the "optimum" ratio of TiB_2 to $AlMgB_{14}$ is described in Section 1.1. Additionally, the effect of variations in composite microstructure on mechanical properties was also examined, as described in Section 1.2.

Another key step toward commercialization was developing a cost-effective method to improve fracture toughness in $AlMgB_{14}$-TiB_2 composites. The indentation toughness (as measured by cracks from a 1 kg Vickers impression) of nominally single-phase $AlMgB_{14}$ is approximately 3 $MPa\sqrt{m}$, while that of the mixed-phase composite ranges from 4 to 5 $MPa\sqrt{m}$. While these values are typical of engineered ceramics, the project sought to raise fracture toughness to 7 to 8 $MPa\sqrt{m}$ without significantly lowering hardness. This work to raise fracture toughness is described in Section 1.3.

Oxygen is an undesirable contaminant that, when present during powder processing, leads to formation of $MgAl_2O_4$ (spinel) in the boride nanocomposite. This is particularly problematic in large-scale processing as control of oxygen becomes increasingly difficult and costly as the size of the handling and processing environments increases. Analysis of consolidated compacts prepared at Ames Laboratory and by NewTech Ceramics reveal spinel contents ranging from a few percent to over 10 percent by volume. We have found that such levels of spinel can significantly reduce the hardness and wear resistance of the material; consequently. Section 1.4 describes investigation of cost-effective solutions to reduce or eliminate the spinel phase in boride nanocomposite compacts.

1.1: Optimization of TiB_2 - $AlMgB_{14}$ Ratio

A major goal of this project was determinations of the $AlMgB_{14}$:TiB_2 ratio that optimizes the composite's wear resistance and low coefficient of sliding friction. During the first part of the *project, the best-performing composition was sought. The metric for determination of a "best composition" emphasized the erosive wear performance, abrasive wear performance, and the coefficient of friction; less emphasis was placed on achieving high microhardness values. Once this composition was determined, it became the "standard material" for use in all subsequent phases of the project.*

The purpose of this task was to find the composition range of mixed-phase boride composites possessing the best combination of hardness, fracture toughness, and wear resistance. While these qualities are not independent, their interrelationship is complex and depends on the particular application (e.g., high-speed machining, erosive or abrasive wear, or the presence or absence of protective lubrication). The most wear-resistant composition known at the start of this project was 30 wt% $AlMgB_{14}$ - 70 wt% TiB_2 (~60 vol% TiB_2) [Ahmed06]. However, the optimum composition had not been conclusively established to within a few percent TiB_2. Moreover, the strong interrelationship between processing and properties in these materials demanded the establishment of strict guidelines for phase size and distribution, porosity, and impurity phases in order to permit meaningful comparisons between samples of various compositions.

In the early years of study of the $AlMgB_{14}$ family of materials, microhardness testing was used as a key performance metric. However, in this project, it was found that erosion rate measurements are more reproducible and less prone to both statistical and human errors than hardness testing. In this study evaluation of samples often involved erosion testing. Unlike hardness testing, erosion is a performance-based test. With the acquisition of an ASTM erosion testing system, it was found that erosion testing was a more useful determination of wear resistance than is microhardness, and erosion testing does not require the time-consuming sample polishing necessary for hardness testing.

Reproducible measurement of the sliding wear behavior of materials studied in this phase of the project was provided by a desktop ASTM abrasion test unit. This unit provides complementary data to the existing dry erosion test system, and proved to be a useful indicator of wear behavior under a range of wear-intensive conditions.

To find the optimum composition for wear resistance within the 60 to 70 volume percent TiB_2 range, samples containing 60, 62.5, 65, 67.5, and 70 vol% TiB_2 were synthesized. Erosion testing, which is believed to be a more performance-based test involving less statistical error than hardness testing, was used to compare performances, and no clear trend was evident. Samples were then synthesized containing 72.5, 80, 85, and 90 vol% TiB_2, using current processing technology. Erosion resistance steadily increased with rising TiB_2 content, reaching a maximum at the 80-85 vol% TiB_2 level (Figure 1.1.1). Erosion resistance decreased sharply at 90 vol% TiB_2. These results are consistent with those reported earlier when differences in processing and sintering environments are taken into consideration. The composition offering maximum erosion resistance thus appears to be highly dependent on processing, a point that is critical to successful commercialization in an industrial environment. While early research implied a maximum in erosion resistance near the 60 vol. % TiB_2 composition, the current series of compacts, produced with higher purity boron, provide indications of a refinement in optimal composition range. A minimum erosion rate of 0.38 mm^3/kg, observed in the 85 vol% TiB_2 sample, is considerably lower than that of the previously studied composites. The previous minimum was 0.50 mm^3/kg for a number of 60 vol% TiB_2 composites, most of which contained MA TiB_2, which is typically finer than MM TiB_2.

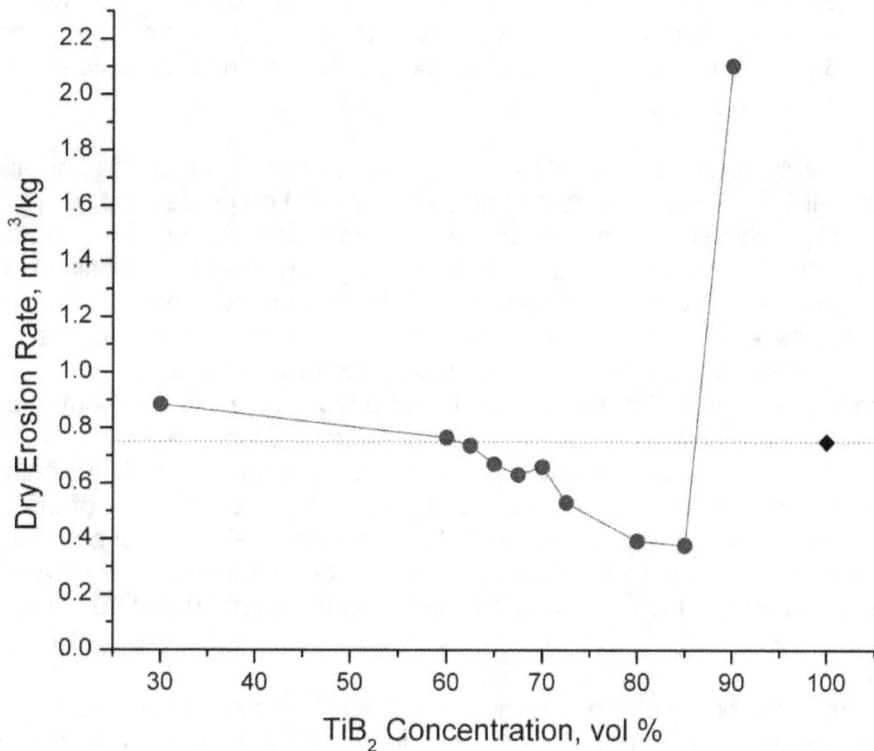

Figure 1.1.1: Erosion rates of $AlMgB_{14}$-TiB_2 composites with varying fractions of mechanically milled TiB_2. The diamond figure represents best erosion resistance achieved for 100 % TiB_2 produced by MM and sintered at 1400 °C. This is coincidentally 0.75 mm^3/kg, the project's stated goal for erosion rate represented by the horizontal line.

Interestingly, the erosion rate of a 100% TiB_2 sample produced under similar conditions is lower than that for the 90 vol% sample, although still not as good as for the $AlMgB_{14}$-TiB_2 mixed phase composites above 60 vol%. Typically, TiB_2 and TiB_2-rich composites are exceedingly difficult to sinter at temperatures less than 1800 °C. Previous research has shown that as the composition approaches 100 % TiB_2, powder processing parameters such as grain size and impurities become more critical to sinterability. Advancements in processing technology developed as a result of this program have enabled improved synthesis of boride-based materials to the extent that the maximum TiB_2 concentration amenable to sintering at 1400 °C has increased from ~60 to ~85 vol%.

Figure 1.1.1 shows equivalent erosion resistance for the 60 and 100 vol% TiB_2 concentrations. Above 60 vol% TiB_2 there is not just greater than rule-of-mixtures performance, but greater than that of pure TiB_2 prepared by nearly identical methods. This again implies a synergy between the two ceramic phases and the importance of $AlMgB_{14}$ as both a sintering aid and reinforcement phase. Yet at 90 vol% TiB_2, wear has drastically increased, this may be due to competing effects of sintering and/or strengthening mechanisms. In the (near) pure TiB_2 samples, it was found that small amounts (1-3 %) of C and N impurities aided in sintering of these samples [Peters09]. It is possible that at 90 vol% TiB_2, small amounts of $AlMgB_{14}$ and Ti(C,N) react unfavorably to cause an overall reduction in sinterability or reinforcement with respect to the use of only one sintering aid. Yet as can be seen in the figure, $AlMgB_{14}$ additions can result in better erosion resistance than Ti(C,N) additions, albeit at significantly different concentrations. As has been reported previously [Peters09], the small additions of Ti(C,N) to

TiB$_2$ aid in sintering to high (99%) densities yet offer little reinforcement as residual sintering aids are associated with intergranular fracture. Thus, while AlMgB$_{14}$ and Ti(C,N) are both good sintering aids for TiB$_2$, AlMgB$_{14}$ offers good bonding and subsequent mechanical reinforcement to the composite.

At the highest levels of TiB$_2$, AlMgB$_{14}$ serves a dual role as a hard reinforcement phase and also as a sintering aid. The optimal TiB$_2$ concentration is strongly dependent on processing parameters, boron purity, and particle size. The unique synergy, in terms of strong interphase bonding, between AlMgB$_{14}$ and TiB$_2$ extends the range of technologically useful compositions to nearly 90 vol. % TiB$_2$. The data in Figure 1.1.1 indicate that while both phases provide intrinsic erosion resistance in the composite, AlMgB$_{14}$ acts in addition to facilitate densification of the TiB$_2$-rich compositions so that wear resistance is not hindered by excessive porosity. Past experience with the composites and pure TiB$_2$ has shown that the higher the TiB$_2$ concentration, the more critical the processing parameters become (i.e. particle size and oxygen contamination) to densification. It should be noted that commercial sintering of TiB$_2$ is typically performed at temperatures exceeding 1800 ºC. As a sintering aid on the TiB$_2$-rich end of the composition range, AlMgB$_{14}$ enables nearly complete densification, finer grain size (of both TiB$_2$ and AlMgB$_{14}$), and fewer impurities than any other addition that has been examined. Past and current experiments that have identified the maximum TiB$_2$ concentrations of 60 and 85 vol%, respectively, support this.

In addition to compositional effects, one major processing effect was investigated in tandem with the above experiment. Figure 1.1.2 summarizes the results of ASTM dry erosion testing on compacts hot pressed at 1400ºC with varying compositions of TiB$_2$ as produced from conditioned vs. unconditioned milling. When milling in a clean steel vial with unused media, the milling is said to be *unconditioned*. After use, the vial and media become encrusted with milled powder. At this point, the vial is said to be *conditioned*, and further milling in the same vial proceeds with steady-state wear mechanisms. Unconditioned milling, discussed above, results in a minimum erosion rate of 0.38 mm^3/kg at a composition of 85 vol% TiB$_2$, whereas a lower erosion rate (0.24 mm^3/kg) was measured for samples prepared in conditioned vessels at a lower TiB$_2$ content (80 vol%).

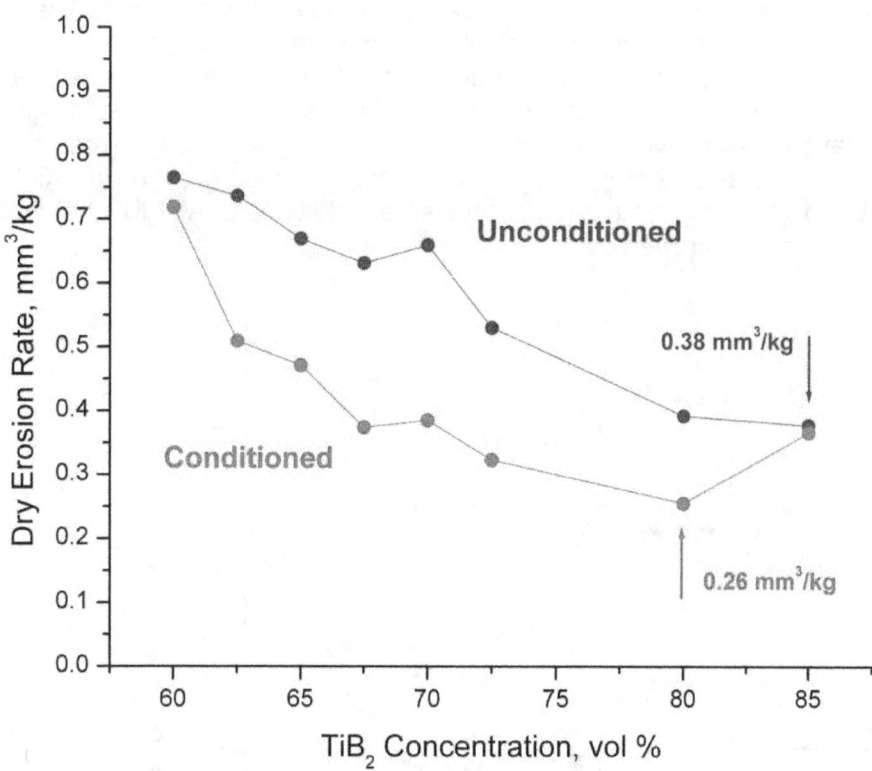

Figure 1.1.2: Erosion results for samples produced by unconditioned and conditioned milling. Unconditioned material was milled in a clean vial, conditioned material was milled in a used and powder-encrusted vial. Conditioning shows greater than 30% improvement at most compositions while the "optimum" composition also shifts.

Density measurements for all samples show a rough trend that reaches a maximum around 75 vol% TiB_2. The erosion resistance of the composites thus appears to increase with TiB_2 fraction up to a maximum concentration of about 90%, even as sinterability decreases slightly. The erosion resistance of conditioned material reaches a maximum at a lower TiB_2 concentration than in the case of unconditioned milling. Conditioned milling appears to result in reduced sinterability along with a more favorable microstructure for wear resistance. The primary factors associated with the different conditions are Fe content of the powders and grain size of the consolidated articles. The Fe content may be higher in unconditioned powder, increasing sinterability while the higher amount of residual Fe lowers hardness after consolidation, which lowers the composite's wear resistance. The higher erosion resistance of the conditioned material may be partially attributable to its reduced Fe content. The lower sinterability of the conditioned material causes increased porosity, which would be expected to degrade erosive wear resistance, but this may be outweighed by the benefit of having a significantly lower FeB impurity content.

As discussed above, first-generation $AlMgB_{14}$-TiB_2 composites exhibited a maximum in wear resistance around the 60 vol% TiB_2 composition [Ahmed09,Peters07]. This is believed to be due to in part by the higher activation energy for densification as the composition approached 100% TiB_2. As shown in Figure 1.1.3, density reached a local maximum at 72.5 % TiB_2, yet the highest erosion rate was seen for the 80% sample. This illustrates that erosion resistance in these composites can be seen as a function of the competing effects of density and TiB_2 concentration. TiB_2 is the more erosion resistant component of the composite, yet pure TiB_2 is difficult to densify. Porosity is strongly detrimental to wear resistance; thus, a compromise must be found between maximum TiB_2 phase fraction (high intrinsic wear resistance) and minimum porosity (by addition of sintering aid). As mentioned earlier, the 100% TiB_2 samples include $Ti(C,N)$ impurities that aid in densification, contributing to the high density seen in Fig. 1.1.3. Yet these sintering aids have relatively weak bonding, resulting in a higher relative erosion rate. Poorly bonding inclusions have a detrimental effect similar to the presence of porosity. $AlMgB_{14}$ on the other hand is a strongly bonding sintering aid, thus small additions of $AlMgB_{14}$ have little effect on TiB_2's intrinsic mechanical properties.

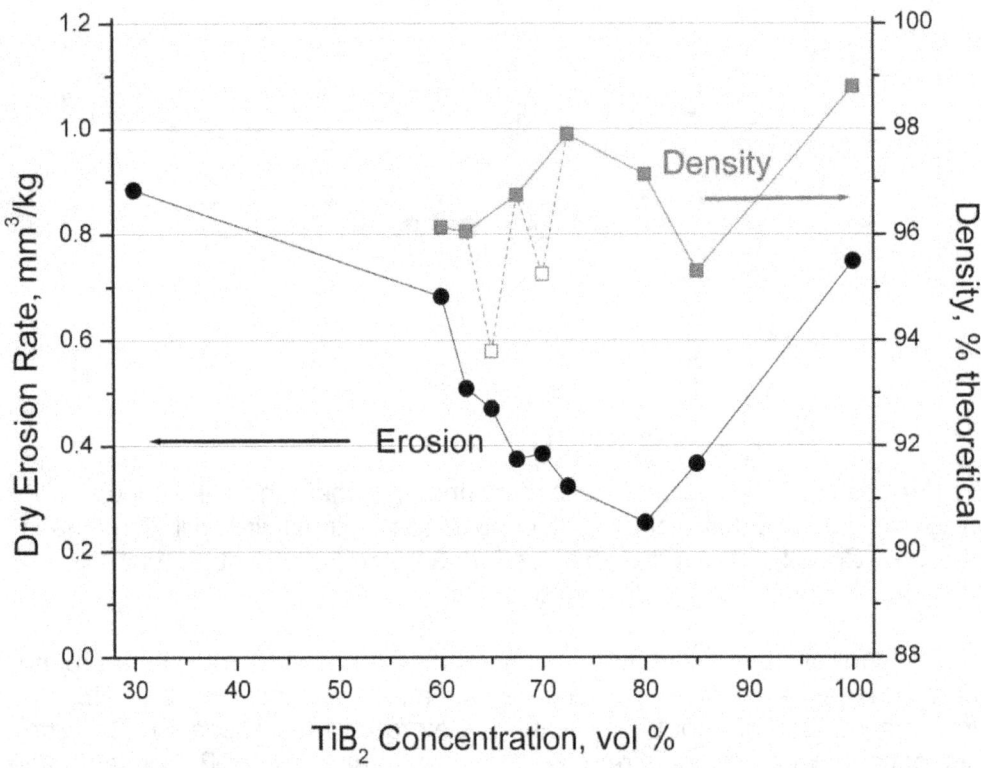

Figure 1.1.3: Density for samples with varying TiB_2, open squares correspond to anomalously low density samples. Erosion rates for comparison of density and behavior; i.e. the low density of the samples indicated by open squares corresponds to slight increases in relative wear rate.

Improvements in processing, namely reduced grain size and impurity concentrations, have improved sinterability of the powders up to 85 vol% TiB_2. This in turn has led to improved mechanical properties due to the desirable mixture of hardness, toughness, and erosion resistance of the composite. Because pure TiB_2 is typically difficult to densify below 1800 °C, the presence of $AlMgB_{14}$ and small amounts of Fe in the $AlMgB_{14}$ + TiB_2 composites act as sintering aids to lower consolidation temperatures to around 1400 °C. The documented synergy between $AlMgB_{14}$ and TiB_2 resulting from strong bonding between the two phases produces an enhancement in wear resistance beyond that of either single phase constituent. As was previously established in the case of unconditioned processing, the minimum erosion rate of 0.38 mm^3/kg was observed at 85 vol% TiB_2. In the case of conditioned milling, a minimum erosion rate for the $AlMgB_{14}$-TiB_2 composites of 0.25 mm^3/kg was observed at a composition of 80 vol % TiB_2. This erosion rate is far lower than the project goal of 0.75 mm^3/kg.

Erosion tests on the conditioned samples clearly suggest an improved erosion resistance compared with samples prepared from unconditioned powder. The microstructures of a few selected samples were compared in order to identify any variation in comminution and densification, as shown in Figure 1.1.4.

Figure 1.1.4: Micrographs of 62.5 and 70 vol% TiB_2 composites consolidated from unconditioned (left) and conditioned powder (right).

Unconditioned milling was also found to yield a finer microstructure with fewer large particles. This might be expected as a result of more aggressive milling without the presence of a powder cushion on the media and vessel walls after conditioning. Also, if encrusted material within the vial and on the media in the conditioned state is included in media/powder ratio calculations, the conditioned milling has a lower media/powder ratio and thus milling could be assumed to be slightly less aggressive from this perspective. Figure 1.1.5 shows larger TiB_2 grains corresponding to the conditioned material.

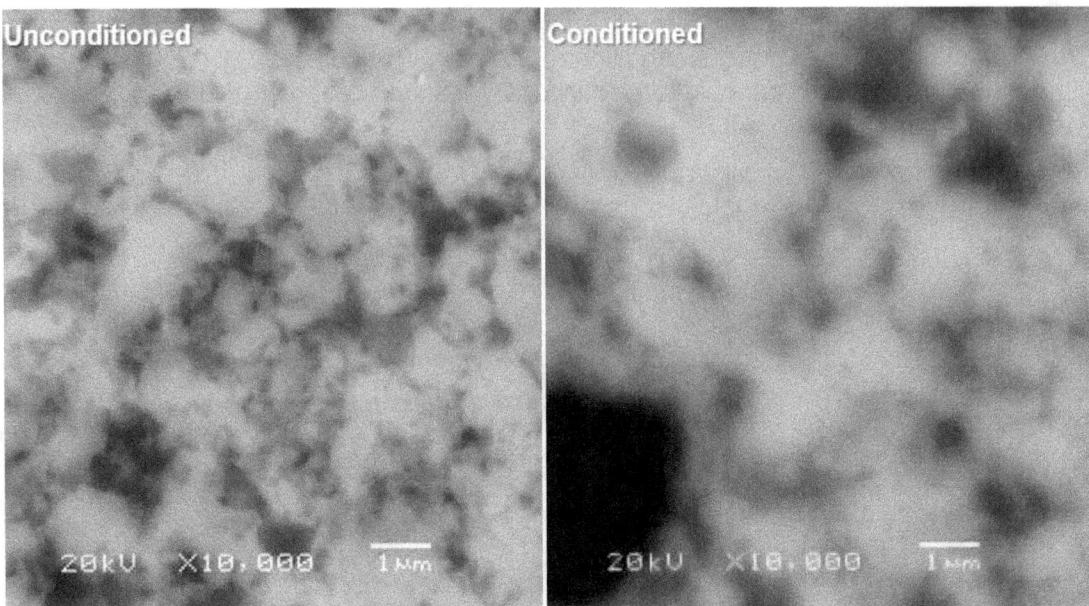

Figure 1.1.5: Backscattered SEM micrographs of 62.5 vol% TiB_2 composites consolidated from unconditioned (left) and conditioned (right) powder.

In both figures, the grains in the conditioned material are larger, which is in agreement with the assumptions of less aggressive milling, yet at odds with expectations of erosion resistance. Typically, fine grain size is associated with improved erosion resistance, the opposite is seen in these materials. This is likely due to an impurity effect; less aggressive milling of the conditioned material results in lower Fe impurities. Like $AlMgB_{14}$, Fe acts as a sintering aid, yet its residual phases are much less desirable because they lower strength. This adds another factor to the balance of porosity and TiB_2 concentration discussed above. Since industrial scale powder production will involve different milling energies and techniques than were used in this small study, significant attention will have to be paid to TiB_2 grain size and Fe contamination in determining the sinterability of the powder and subsequently the maximum TiB_2 concentration that will be achievable. In summary, this experiment has shown that the composite's erosion resistance is strongly dependent on the processing, but the aim for any process should be a maximization of TiB_2 content while minimizing porosity.

Optimal TiB_2 Concentration for Abrasion Resistance
As described above, we used ASTM erosion tests to measure the performance of hot-pressed $AlMgB_{14}$-TiB_2 compacts possessing varying ratios of the two boride phases. Erosive wear can be defined as degradation or loss of material resulting from impinging solid particles, generally in a dry (gaseous) environment. In contrast to erosive wear, abrasive wear results from the sliding or rubbing interaction of two solid objects in which the abrasive is entrained in either a solid or liquid host. Since abrasive wear is also an important performance metric for advanced materials, we have examined the same set of samples in ASTM abrasion tests during the current reporting period. The samples examined include the following compositions: $AlMgB_{14}$ – X vol. % TiB_2, where X = 60, 70, 80, and 85. In addition, a sample of fine grained, wear-resistant WC-6%Co was also examined for comparison. The samples were prepared for abrasion testing by using electrical discharge machining to obtain a 3 mm by 9 mm flat surface on an edge of a larger specimen. The flat ensures that all samples have a similar surface area during the tests, so that the number of abrasive particles abrading the sample per unit area and the contact pressure are constant.

The sample is securely held in place against a #400-grit diamond abrasive belt by a series of weights, and the belt is partially immersed in de-ionized water to prevent accumulation of debris on the surface. The long axis of the flat is oriented in the direction of belt travel. The belt is inspected between runs for evidence of wear (loss of diamond grit) and the track of the sample is adjusted between runs so that the test is always started on an unused portion of the belt. Wear is determined by mass loss between runs, during which the sample is wiped with an acetone-soaked cloth, ultrasonically rinsed in methanol, dried with an air gun, and then weighed on a 6-place analytical balance. From the known density of the sample, the volume loss can be calculated. Wear rate is defined as the volume loss per unit distance traveled of the diamond belt (mm^3/m). Figure 1.1.6 shows the abrasive wear results of these samples as a function of load and belt speed. As a comparison, the results of abrasion tests on the tungsten carbide sample are shown in Figure1.1.7. Note that the scale on the y-axis (abrasive wear rate) is an order of magnitude higher for the carbide sample than for the boride samples. Among the four $AlMgB_{14}$-TiB_2 compositions, the differences in abrasive wear rates are insignificant. All samples exhibit a general decrease in abrasive wear with increasing belt speed. At low speeds thermal effects at the interface are negligible and so cutting action by the diamond abrasive particles leads to comparatively high wear rates. With an increase in speed, the temperature at the interface increases so that the effect of the abrasive particles transitions to plowing and gouging which are less efficient wear mechanisms than the cutting occurring at lower speeds. At the highest belt speed, there is relatively little cutting action, so the abrasive wear rate reaches a minimum. Since abrasive wear is a function of hardness and toughness, the similarity between the four boride samples leads to similar wear rates.

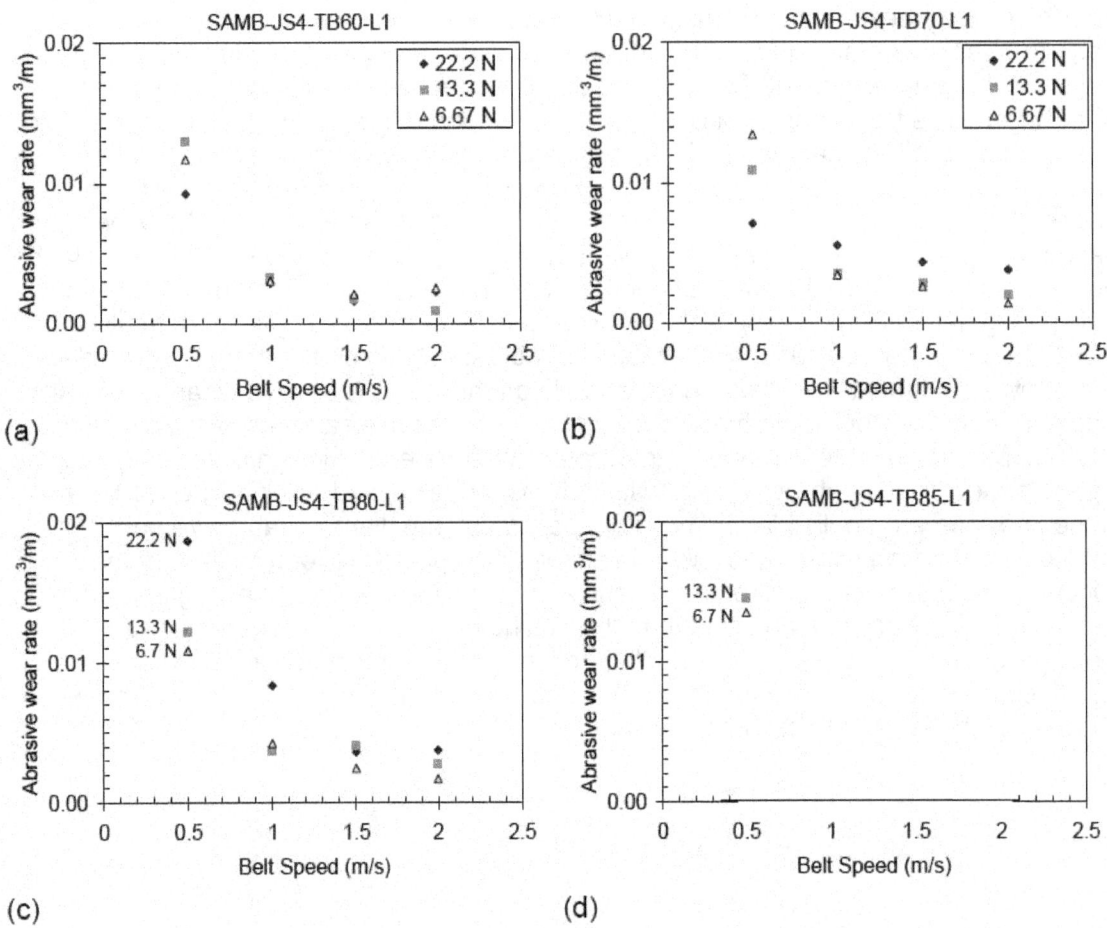

Figure 1.1.6. Results of abrasive wear studies on $AlMgB_{14} - x$ wt. % TiB_2 where x = 60 (panel a), 70 (panel b), 80 (panel c), and 85 (panel d).

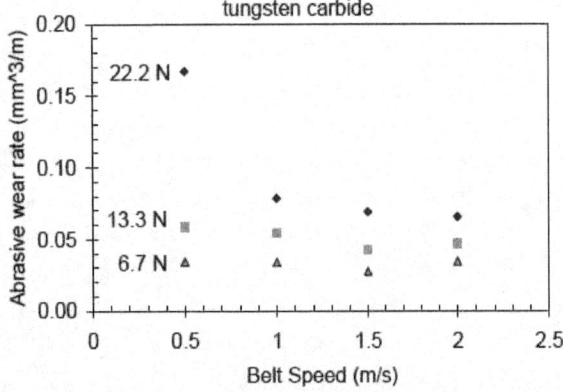

Figure 1.1.7. Results of abrasive wear studies on fine-grained WC-6%Co.

As expected, the abrasive wear rate generally increased with an increase in load. There is some scatter at the lowest belt speed among the boride samples, but this is because the wear rates are so low. The much lower wear rate in the boride materials is a consequence of the higher hardness and lower penetration depth of the diamond abrasive into the material's surface.

1.2: Microstructure

Efforts during the current reporting period to pinpoint the optimum ratio of TiB_2 to $AlMgB_{14}$ have generated a number of useful supplementary findings regarding these materials' microstructures. While pressing-sintering has been identified as the most cost-effective processing approach for industrial scale-up activity (see Task 2, below), a number of the lab-scale studies that fall under Task 1 are most efficiently carried out by mechanical milling (MM) and/or mechanical alloying (MA) in 200 cm^3 Spex vials. The advantage of this approach is that compositional variables can be quickly screened for their effect on wear resistance, and the results of these studies can be subsequently incorporated into the larger-scale pressing and sintering efforts. Discussed in Section 1.1 above, microstructure can be as important as composition with respect to mechanical properties. Correlations between microstructure and erosion of samples produced by MM and MA were studied previously. In summary, MA typically resulted in higher erosion resistance which was attributed to finer grain size of the TiB_2 constituent.

In-situ TiB_2 Processing

While investigating the optimum composition of these composites, a novel processing technique referred to as the "in-situ" method of TiB_2 addition has generated some particularly interesting results. This method consists of simultaneously milling Al, Mg, Ti, and B to form TiB_2 by a spontaneous combustion synthesis reaction. This differs from the MM method, which involves mechanical pulverization of commercial TiB_2 powders, and the MA method, in which the TiB_2 phase is formed by alloying of Ti metal and B prior to addition of the $AlMgB_{14}$ powder. The $AlMgB_{14}$ phase is typically formed during consolidation of the elemental powders (called reaction sintering), which are much more easily sintered than pre-reacted $AlMgB_{14}$. Thus, in this discussion, *MM*, *MA*, and *in-situ* refer to the state of the TiB_2 addition. Previous research has shown that this in-situ powder processing approach is capable of producing material with an exceptionally fine microstructure and in many cases, a correspondingly high erosion resistance. Samples of 40, 50, and 60 vol % were produced and evaluated for hardness and erosion resistance. The microstructure of each was also examined by SEM.

The TiB_2 phase was formed during "in-situ" processing wherein all of the elemental constituents are simultaneously subjected to high energy milling, which produces a finer and more uniformly distributed microstructure. In addition, impurity phases accumulated during processing should not affect the $AlMgB_{14}/TiB_2$ ratio, resulting in more accurate relationships between properties and composition. Chunk boron was also used as a starting material to reduce the oxygen normally associated with fine boron powders, which in turn reduces the amount of $MgAl_2O_4$ impurity phase.

Additional samples have been prepared to simulate previous materials which have shown the highest erosion resistance. Theories about the superior performance observed in some previous samples focus on grain size and both Fe and O concentration. Oxygen has been an unavoidable contaminant introduced by the fine B precursor powders. As mentioned earlier, Fe concentration can be reduced by changing milling processes but some Fe is known to be beneficial to sintering. Finer grain size is also achieved by using TiB_2 produced by mechanical alloying (MA) rather than milling (MM) of commercial TiB_2. Some sample microstructures of this are shown in Figure 1.2.1.

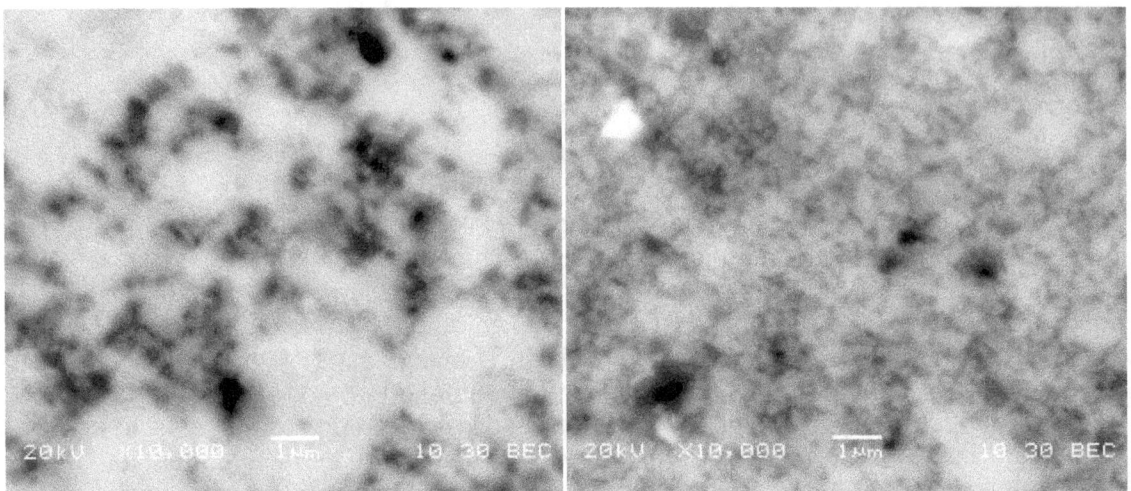

Figure 1.2.1: Electron micrographs of 70 wt% TiB_2 composites containing commercial TiB_2 (left) and mechanically alloyed TiB_2 (right).

The 40 to 60 vol % samples exhibited similar microstructures, consisting of a sub-micron mixture of TiB_2 and $AlMgB_{14}$ (Fig. 1.2.2). There are also larger, faceted crystals that are enriched in boron. These crystals are not observed in composites produced by other methods. The exact structure and composition of the boron rich (dark) phase has not yet been identified, since their small size, combined with a high boron concentration, makes semi-quantitative EDS analysis unreliable. In addition, the strong X-ray scattering power of TiB_2 exacerbates the task of minor phase identification by XRD. Conflicting XRD and EDS results provide ambiguous indications about whether these faceted crystals are $AlMgB_{14}$.

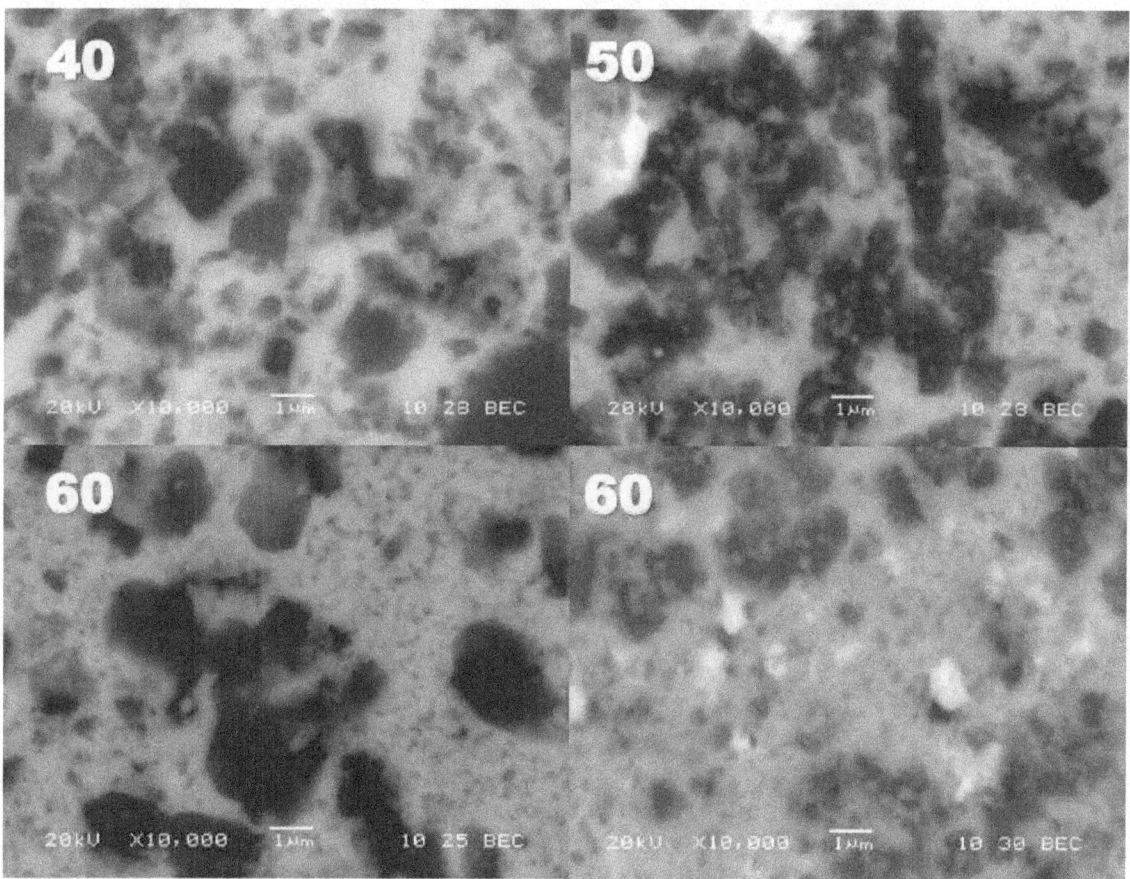

Figure 1.2.2: Electron micrographs of 40, 50, and 60 vol % in-situ samples. Bright regions correspond to Fe, dark regions are B-rich faceted crystals, and grey regions are a fine mixture of TiB_2 and a B-rich phase.

Wear tests of the aforementioned samples showed an increase in erosion resistance with increasing TiB_2 content from 40 to 60 vol % (Figure 1.2.3). These results are within the expected range for these compositions, but do not surpass the best samples produced (i.e., 60 vol% samples with MM TiB_2 can exhibit erosion rates as low as 0.5×10^{-3} mm^3/g (0.5 mm^3/kg). The "in-situ" TiB_2 approach results in differences not only in grain size, but also in impurities, morphology, etc., which affect wear resistance. In addition, the in-situ samples appear to be more susceptible to processing variations, which will be discussed in more detail below. The 60 vol % sample exhibited the lowest erosion, despite the presence of distinct regions showing significantly higher erosion loss than the remainder of the sample. However, indications are that the observed trend is accurate and more significant improvements in erosion resistance between 50 and 60 % may be possible.

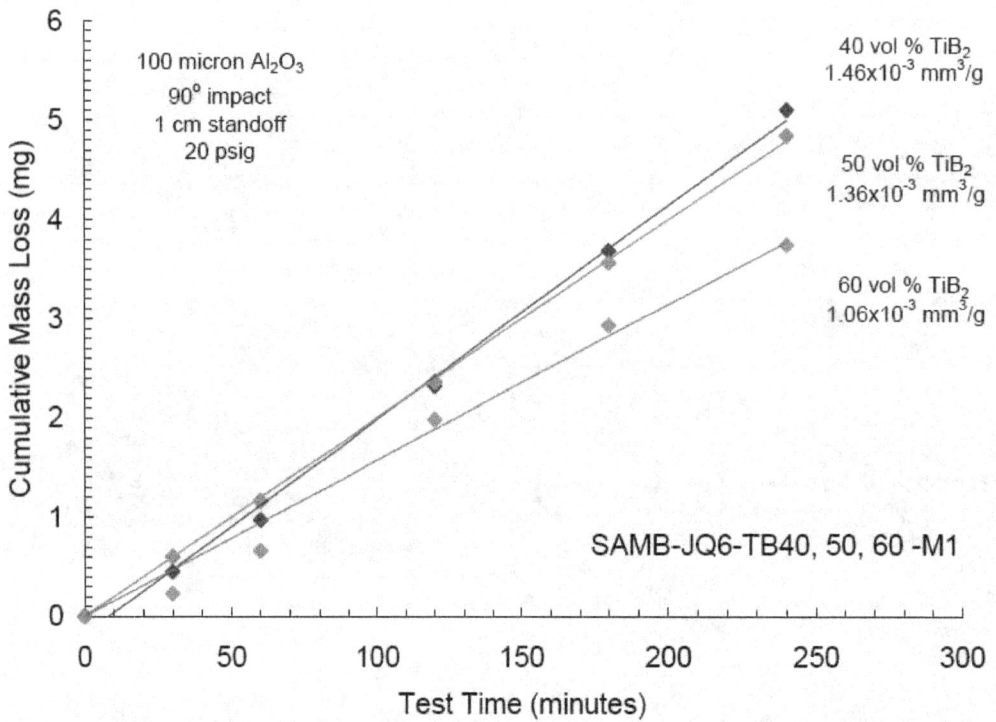

Figure 1.2.3. Results of steady-state erosion tests on samples of AlMgB$_{14}$ containing 40, 50, and 60 volume percent TiB$_2$ prepared by the "in-situ" method.

It has also been discovered that the prior conditioning (§1.1) of the milling vessel can affect the microstructure and as a consequence, the erosion resistance. For example, the 60 vol % sample pressed from powder generated by the first (or un-conditioned) milling exhibited a similar microstructure to the 40 and 50 vol % samples discussed previously (Figure 1.2.2). Subsequent milling runs in the conditioned vial resulted in microstructure variations, as shown in Figure 1.2.3. The microstructures resulting from the second and third milling runs consist of a fine and evenly distributed mixture of TiB$_2$ and an unidentified low atomic number phase. XRD indicated that the amount of crystallized AlMgB$_{14}$ phase was below the 5 vol % limit. After the third milling run, the resulting powder had also accumulated a significant amount of Fe contamination (bright phase), not observed in samples prepared from the first and second milling. As the milling container becomes conditioned, a small amount of residue from previous runs accumulates in corners, near the top and bottom ends of the vial. Due to the excessive Fe contamination of the third run, the milling could be said to be "over-conditioned." This effect was not seen in extended milling of either the MM or MA samples, once again demonstrating the sensitivity of the in-situ method to processing conditions.

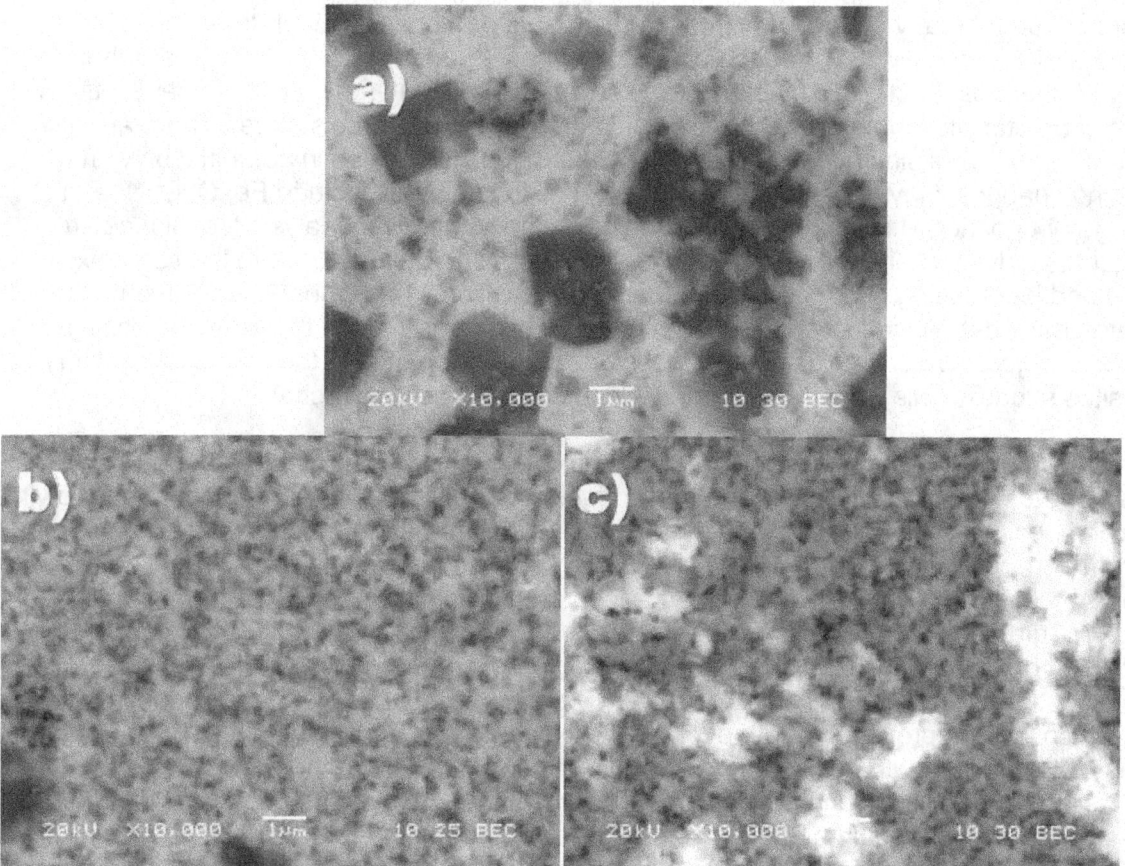

Figure 1.2.4: Micrographs a, b, and c of 60 vol % samples pressed from powder from first, second, and third milling runs, respectively, showing the effect of conditioning of the milling vessel. The erosion rate of sample b) was less than for sample a); the erosion rate of sample c) was quite high, presumably due to excessive Fe contamination.

Erosion tests have shown an unambiguous relationship between grain size and erosion resistance. Finer microstructures generally exhibit marked improvement over coarser microstructures resulting from mechanical mixing of commercial TiB_2 powders. As shown in Figure 1.2.4, milling in a conditioned vial produces fine, evenly distributed microstructures with features on the order of 200 nm. The relative erosion rates of the first, second, and third millings were 1.24, 1.01, and 3.75 mm^3/kg, respectively. The microstructure in Figure 3c eroded significantly faster due to high Fe pick-up. However, XRD indicates a lower volume fraction of $AlMgB_{14}$ in these in-situ composites, suggesting that the dark features may be off-stoichiometry phases, which are not as strong or as compatible with TiB_2 in terms of interfacial bonding as $AlMgB_{14}$. Due to the small regions that were sampled during the tests, these observations are only qualitative. These results are important in that they help to define the relationship between processing options and expected wear resistance, and for down-selection of the best approach for industrial-scale production. In-situ processing shows potential to generate fine, even microstructures, yet appears to be susceptible to contamination and/or other processing parameters.

Retaining the fine distribution of TiB_2 found in the "in-situ" composites is expected to provide substantial improvement in wear resistance. There is some evidence that the B-rich phase in the in-situ composites may not be entirely $AlMgB_{14}$. As we reported in a recent journal article, TiB_2 and $AlMgB_{14}$ are highly compatible, strongly bonding phases [Cook10]. This is not necessarily true for many other B-rich phases (high borides), possibly even those possessing the same structure as $AlMgB_{14}$. If such phases are present, many would be less hard than the $AlMgB_{14}$ phase (based on our previous characterization of AlB_{12}) and would also have different bonding characteristics, both of which would ultimately affect wear resistance. There are also many known and possible impurities present in the composites, which in combination with processing changes, may generate new phases. These impurities include Fe, O, C, N, and Cr. Due to the abundance of light elements B, O, and C, most chemical analysis techniques are insufficient to solve this problem. Additionally, XRD analysis is difficult due to the complex patterns and low scatting potential of high boride compositions. The most effective route to complete phase determination would be in-depth TEM analysis. While this was not feasible in the current project, evidence collected on possible unknown phases in the in-situ (and other) composites suggests this would be a worthwhile topic for future research.

Pure, monolithic TiB_2 has a reputation for being difficult to sinter. Sintering temperatures as high as 1800°C were typically used, and these high temperatures caused rampant grain growth, producing specimens with a wide dispersion of grain sizes and average grain sizes in the tens of micrometers or larger. The pure, monolithic TiB_2 samples produced in this project were sintered successfully at only 1450°C and had an average grain size of around 1 μm or greater. Yet the TiB_2 grains in the in-situ composite samples fall between 100 and 500 nm in diameter, showing that grain growth is strongly suppressed by grains of the $AlMgB_{14}$ phase. Fracture comparisons between pure TiB_2 and $AlMgB_{14}$-TiB_2 composites show that phase boundaries with $AlMgB_{14}$ also have improved strength over that of single-phase TiB_2. While $AlMgB_{14}$ appears to suppress TiB_2 grain growth in all of the composites, the distribution of the $AlMgB_{14}$ phase is quite different in the in-situ composites compared with those prepared from MA or MM TiB_2. The in-situ composites required a lower volume fraction of $AlMgB_{14}$ to constrain the TiB_2 grains than is needed to do the same in the MM and MA composites. Figure 1.2.5 shows an in-situ and MM composite side by side. Ignoring the large dark crystals, $AlMgB_{14}$ more efficiently pins the TiB_2 particle boundaries in the in-situ composites, thus preventing grain growth and strengthening grain boundaries at lower concentrations. Only a small fraction of the $AlMgB_{14}$ in the in-situ material is pinning grain boundaries, the rest is "wasted" in forming large boron-rich crystals. As these two composites have the same nominal fraction of $AlMgB_{14}$, the fraction necessary in an in-situ composite is lower and at the same time generates a finer microstructure.

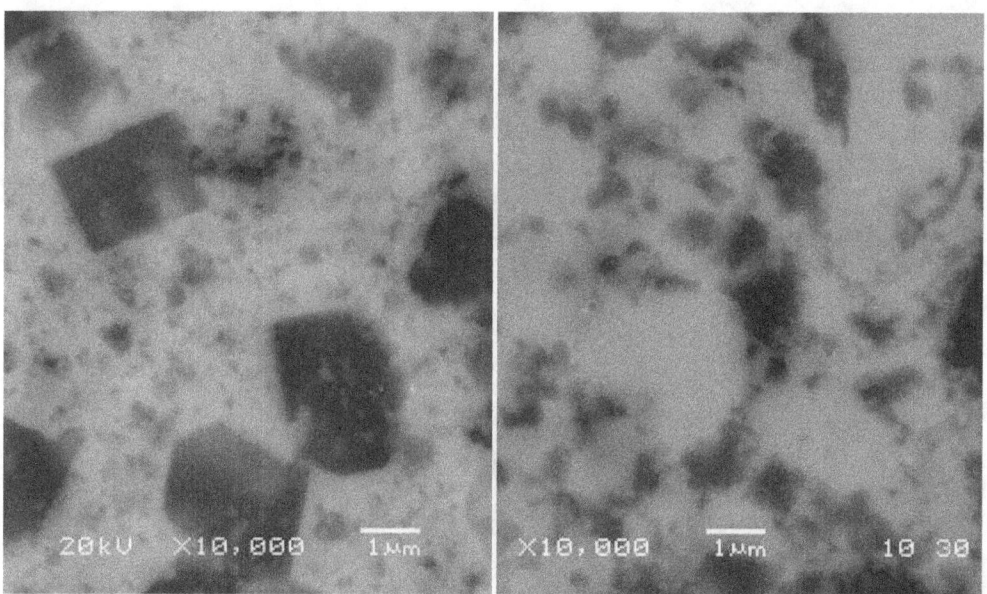

Figure 1.2.5: Left: 60 vol% in-situ TiB$_2$ composite microstructure containing large crystals and fine TiB$_2$ distribution. Right: 60 vol% MM TiB$_2$ composite containing coarser TiB$_2$.

Minor amounts of phases such as AlB$_{12}$ have been suspected in certain samples but are difficult to detect due to their complex XRD patterns and the fine grain size of the composites. If such phases are present, they would possess lower hardness than the AlMgB$_{14}$ phase and would also have different bonding characteristics. Figure 1.2.6 gives examples of some unusual microstructural features that were observed only in compacts produced by unconditioned milling. These were not seen in any of the conditioned samples, yet due to differences in image quality, this finding does not disprove their existence. The regions of a fine mixture of TiB$_2$ and dark B-rich particles are seen in all three micrographs in the figure. If the AlMgB$_{14}$ is stable, this is a highly desirable microstructure. If another phase is present, bonding is likely poor and wear resistance would be lower.

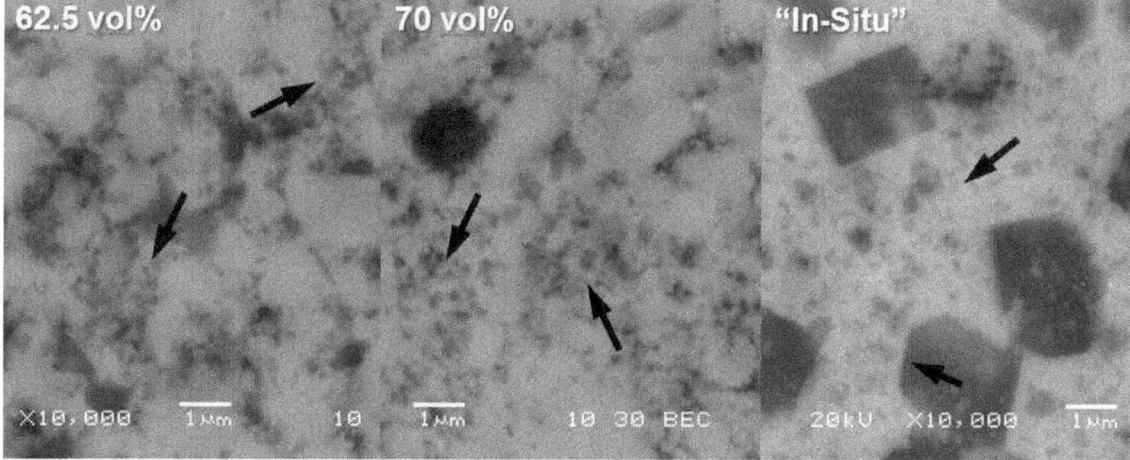

Figure 1.2.6: Unconditioned composite micrographs compared to a composite produced by "in-situ" milling of AlMgB$_{14}$ and TiB$_2$. The arrows indicate similar features that in the case of the "in-situ" materials were not believed to be AlMgB$_{14}$ and appeared to lead to higher erosion rates.

Grain Size Effects

As discussed above, grain size plays an important role in erosion resistance; typically finer grains yield higher resistance. This difference has been seen most often in comparison of the MM and MA composites. While the MA composites exhibit an average grain size. In order to verify this trend over a wider range, large-grained (10-100 µm) composites were produced by the aforementioned "in-situ" method and by the addition of as-received TiB_2 with porosity and impurities concentrations closely matched to previous fine-grained (<1 µm) "in-situ" composites. Both involved short duration milling of the 30 vol% TiB_2 addition to mix the powders with minimal grain size reduction.

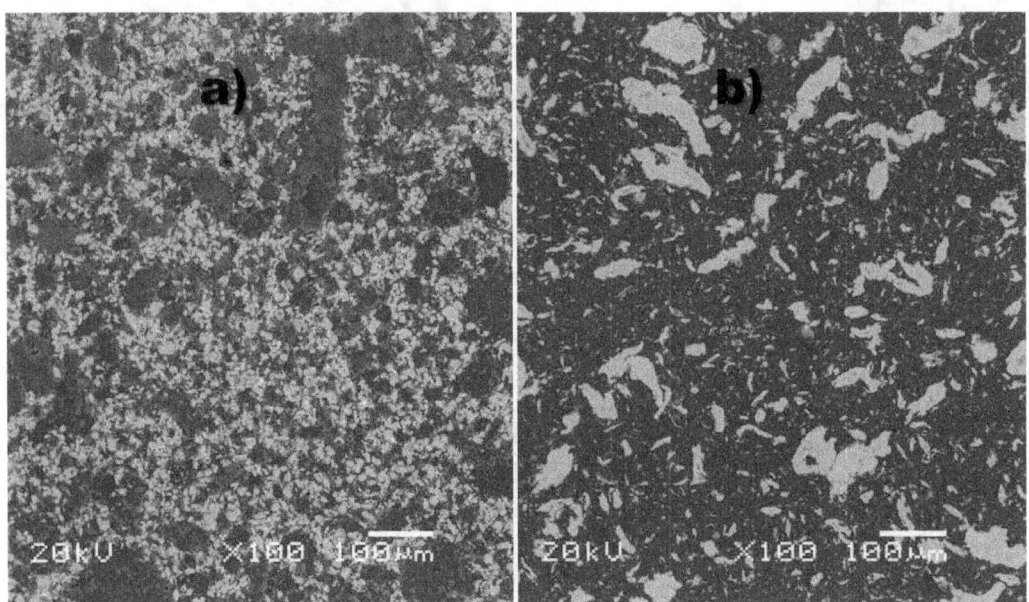

Figure 1.2.7: Two composites containing intentionally large-grained TiB_2 for erosion studies; a) Conventionally-prepared using commercially available TiB_2 resulting in higher porosity, b) In-situ TiB_2 resulting in improved bonding with matrix and lower porosity. The large-grained samples were used to study trends of wear resistance as a function of grain size.

To initiate evaluation of the effect of varying grain size on erosion resistance, a set of three 70 wt. % TiB_2 samples were prepared in such a way that their only difference was the average grain size of the TiB_2 phase. Results of steady-state erosion tests on these three samples are shown in Figure 1.2.8.

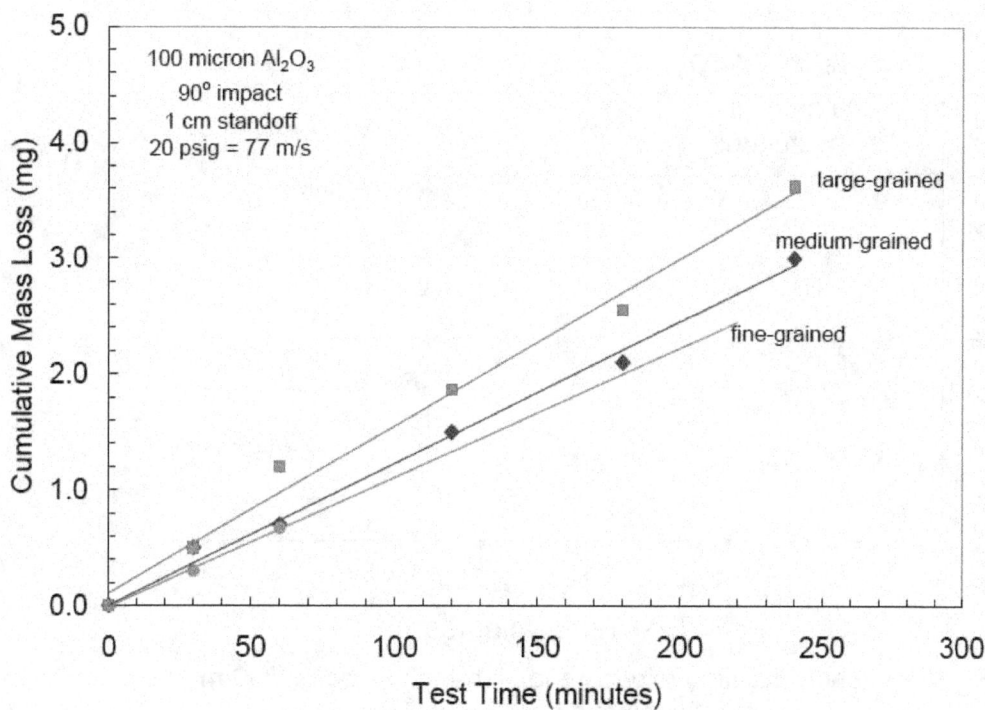

Figure 1.2.8. Steady state erosion for three AlMgB$_{14}$-70 wt. % TiB$_2$ samples with different TiB$_2$ grain sizes. (Erodent impact velocity was 77 m/s).

The steady state erosion rates for the three samples were determined to be 9.4×10^{-4} mm^3/g for the larger (milled), 7.0×10^{-4} mm^3/g for the medium grain-sized (MA Ti+2B) composition, and 5.0×10^{-4} mm^3/g for the fine-grained material, consistent with the general trend of increased erosion resistance with decreasing reinforcement phase size. These tests were performed at normal impact with nominally 100 micron alumina at 77 m/s.

In another set of experiments involving short-duration exposure to a high-velocity (~ 200 m/s) flux of abrasive alumina grit, the erosion behavior of three materials was compared: RocTec 500, monolithic diamond, and a fine-grained AlMgB$_{14}$-70 wt. % TiB$_2$ composite. The purpose of this test was to evaluate the relative volume loss of three of the most technologically significant wear-resistant materials under exceptionally harsh and aggressive conditions, results that will be of interest for future applications in severe-service valves and other particularly demanding environments. Because of the aggressive nature of the test, only short duration exposures were investigated. Results are shown in Figure 1.2.9.

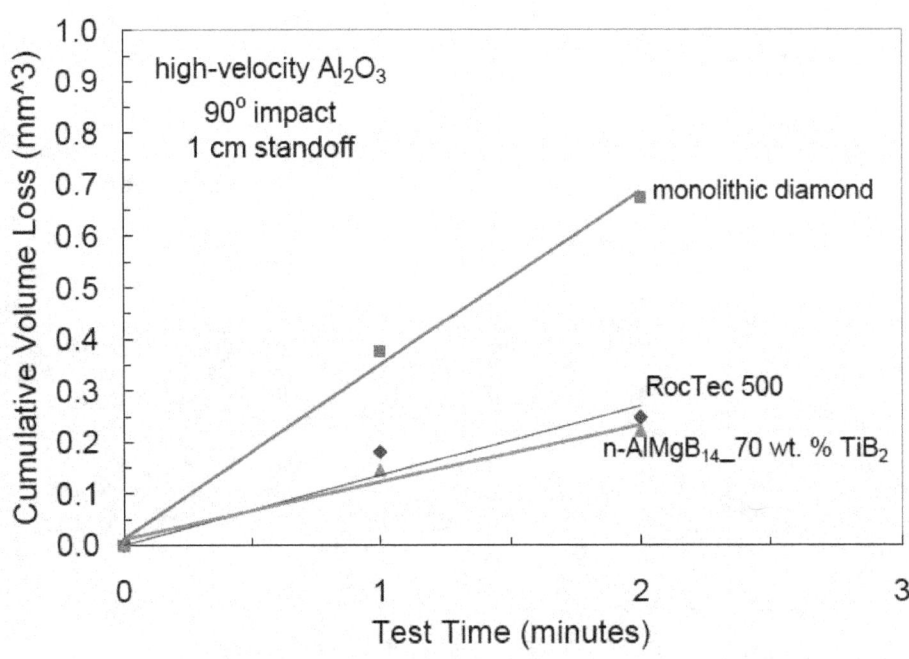

Figure 1.2.9. Results of short-duration exposure to ultra-high velocity (200 m/s) erosion from angular alumina abrasives.

Monolithic diamond exhibited the greatest volume loss, 0.67 mm^3 after 2 minutes, whereas RocTec 500 and the nanoscale boride composite experienced significantly less volumetric losses of 0.25 and 0.22 mm^3, respectively. The superior wear resistance of the boride composite is primarily due to grain size refinement combined with reduced porosity.

Abrasion

Initial calibration of the abrasion test system was conducted on a ½" diameter wear-resistant tungsten carbide insert. The carbide of ISO designation CNMG120408 was a straight WC–6%Co grade with a grain size of 1 µm (Grade 883), supplied by Carboloy-Seco, Inc.

In one series of experiments, the tests are conducted for a combination of 4 different surface speeds (0.5, 1.0, 1.5, and 2.0 m/s) with 3 different loads: 6.67N, 13.34 N, 22.2N, generating 12 independent data points per sample. In another series of experiments, the surface speed is held constant at 1.0 m/s, and the loads are set at 15.56 N and 24.46N, the latter in conjunction with the earlier tests provides abrasion data for 5 loads at a speed of 1.0 m/s. For each test condition, three repeat tests are run. Belt speeds were determined by measurement with a precision stroboscope. Before and after each abrasion run, the specimen was rinsed with acetone and then ethanol, and dried in air. It was weighed in a precision balance to an accuracy of 10^{-5} g in order to determine the loss of material during the abrasion test. Knowing the density of the sample, the abraded volume was calculated from $v = \Delta m/d$, where Δm is the mass loss and d is the density. The total distance traveled was determined from the product of belt speed (in m/s) and time (in s). Abrasive wear rate is given by the quotient of total abraded volume divided by total distance traveled (mm^3/m).

Results for the WC *de facto* standard are compared with a sample of AlMgB$_{14}$-70 wt. % TiB$_2$ (SAMB-TB70-TC2-M3) for three different loads in Figure 1.2.10.

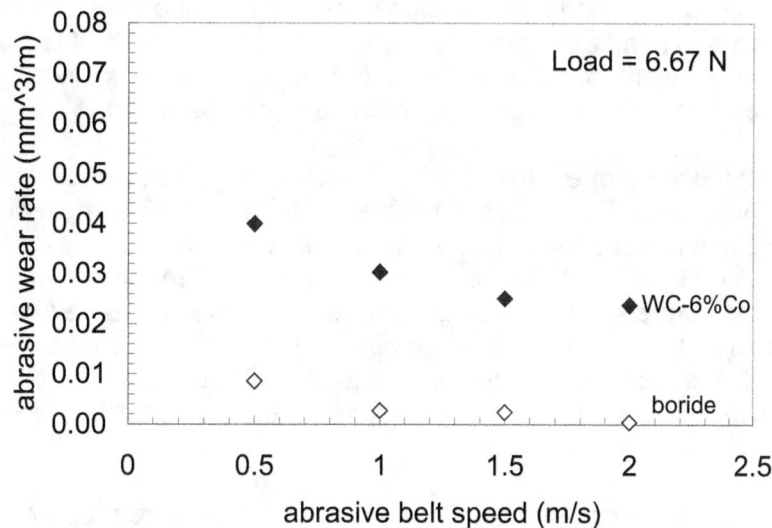

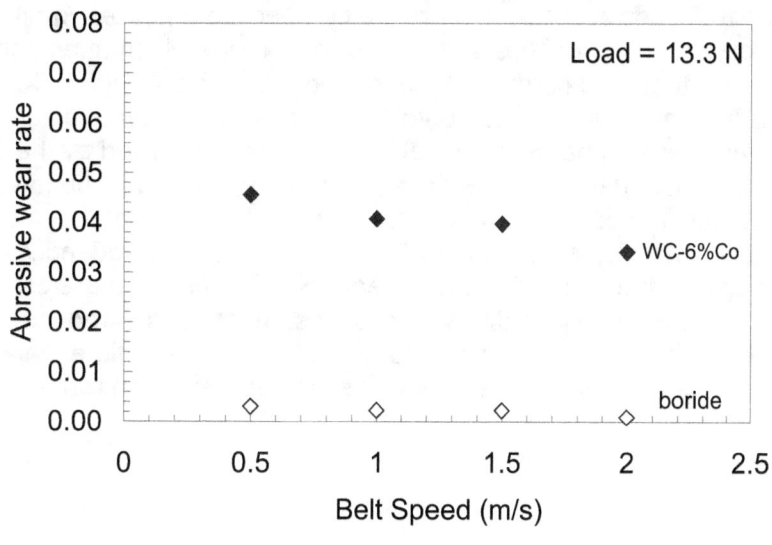

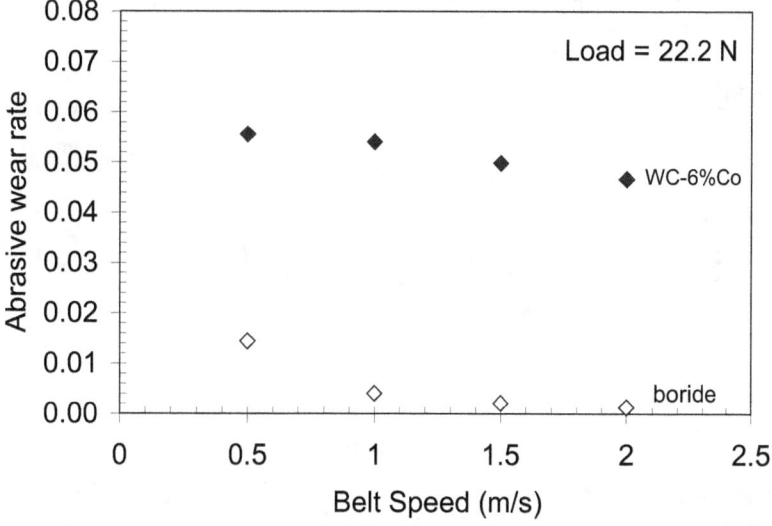

Figure 1.2.10. Results of diamond belt abrasion tests at three different loads on fine grained WC-6%Co (filled symbols) and a hot pressed sample of $AlMgB_{14}$ - 60 vol % MA TiB_2 (open symbols). These results were for initial testing of the abrasion unit early in the project; the boride sample tested was a then-current state of the art composite.

It was observed that the wear rate exhibits a general tendency to decrease with an increase in belt speed for any given load. For example, the wear rate of the WC sample decreases from 0.056 mm^3/m at 0.5 m/s under a load of 22.2N to 0.047 mm^3/m at 2 m/s, a decrease of 16%. In contrast, the wear rate of the boride sample decreases from 0.014 mm^3/m at 0.5 m/s to < 0.001 mm^3/m at 2 m/s. The boride sample exhibited consistently lower abrasive wear rate than the carbide, by at least a factor of four. In fact, the abrasive wear rate of the boride composite at the highest speed, 2 m/s, was too low to measure at the 6.7 and 13.3 N load levels. These results are qualitatively consistent with those obtained previously using different test equipment [Ahmed09].

In one set of experiments, samples of 50, 60, and 70 vol% TiB_2 were prepared with Ti + B milled 6 hrs (MA TiB_2) and $AlMgB_{14}$ milled less than 6 hrs. The shorter milling time was applied in order to reduce the total amount of Fe contamination in the powder. However, as a consequence of the reduced Fe concentration, the compacts did not reach full density during sintering, and all were found to possess some residual porosity which affects the wear rate. Results of erosive wear tests are compared to a previous sample in Figure 1.2.11. A noticeable increase in erosion resistance from 50 to 60 vol% is observed, while the difference between the 60 and 70 vol% samples is comparatively smaller and partially masked by the scatter in mass loss, which is likely a consequence of uneven distribution of porosity. The lower erosion rate of the MM TiB_2 sample designated "TC2" is due to its reduced porosity and standard processing. Fe is an important sintering aid; levels too low have noticeable consequences on wear properties. Based on this study and Task 1.1, ideal Fe concentrations are estimated to be between 2 and 4.5 wt% Fe. In MM and MA composites, most Fe is introduced in the $AlMgB_{14}$ milling, so Fe concentration can vary depending on $AlMgB_{14}$ - TiB_2 ratio even with the same processing. Better Fe distribution will also minimize the concentration required.

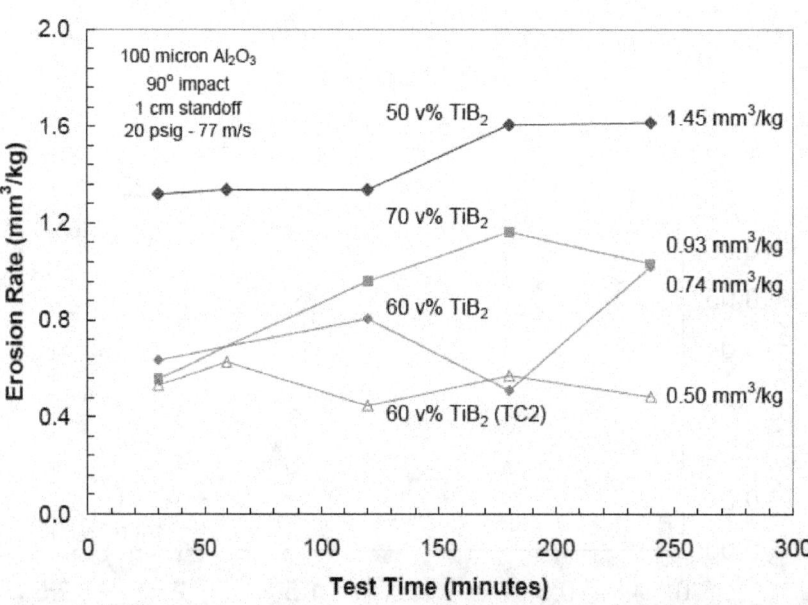

Figure 1.2.11. Segment erosion rates for three composites with MA TiB_2 compared with past sample containing MM TiB_2 (bottom). Average erosion rates are listed to the right.

1.3: 2nd and 3rd Phase Additions

In much previous work, it was established that TiB_2 was a viable reinforcement phase for $AlMgB_{14}$-based ceramic composites [Lewis01,Cook00,Peters07]. Because of the impressive wear resistance observed in the composites, the question arises as to whether further increases in wear resistance could be achieved through the incorporation of secondary reinforcement phases. Given that the Group IVB elements share similar properties, it is important to compare the effect of Zr and/or Hf diboride additions with that of TiB_2, and to optimize the best combination of additions for maximum wear resistance. In addition, many transition metals form diborides with the same structure as TiB_2 (hP_3) and share many of the desirable mechanical and chemical properties. These diborides exhibit either partial or complete solid-solubility with TiB_2, which could result in some novel microstructures, including precipitation of ultrafine reinforcement particles. Consequently, we investigated boride compositions containing alternate mixed diboride reinforcement phases.

CrB_2, ZrB_2, and HfB_2 were chosen for their similarities to TiB_2 and for their resistance to oxidative, corrosive, or other types of degradation. These mixed diboride phases consist of $(Zr,Ti)B_2$, $(Hf,Ti)B_2$, $(Cr,Ti)B_2$; each a 50:50 mixture of the individual diborides. ZrB_2 and HfB_2 were added from commercial powder, similar to the TiB_2 currently employed in these studies. The samples were prepared identically to the MM TiB_2 reinforced $AlMgB_{14}$ wherein the reinforcement phase was added to mechanically alloyed $AlMgB_{14}$ via a 30-minute milling cycle. The samples were all hot pressed at 1400°C for 60 minutes under 100 MPa applied pressure. The resulting compositions then have the form of 40:30:30 ($AlMgB_{14}$:TiB_2:XB_2) by volume. As the melting points of ZrB_2 (3245 °C) and HfB_2 (3380 °C) are higher than that of TiB_2 (3225 °C), neither of these additions densified as readily as the standard composites, thus additional samples were pressed up to 1500°C. Results, including erosion testing, are listed in Table 1.3.1.

Table 1.3.1: Composition, pressing temperature, resulting density, erosion rate for selected mixed di-boride additions to $AlMgB_{14}$-TiB_2.

Addition	Pressing Temp	Density		Erosion Rate
vol%	°C	g/cm^3	% theoretical	mm^3/kg
30 CrB$_2$ - 30 TiB$_2$	1400	3.956	93.8%	1.13
30 CrB$_2$ - 30 TiB$_2$	1500	4.004	94.9%	1.14
30 ZrB$_2$ - 30 TiB$_2$	1400	3.984	88.5%	NA
30 ZrB$_2$ - 30 TiB$_2$	1500	4.302	95.6%	0.87
30 HfB$_2$ - 30 TiB$_2$	1400	5.667	93.0%	0.71
30 HfB$_2$ - 30 TiB$_2$	1500	5.793	95.0%	1.13
60 TiB$_2$	1400	3.825-3.905	95.7-98.6%	0.76-1.80**

*   % density based on impurity estimates for $AlMgB_{14}$-TiB_2 composites.
** erosion data for samples produced from same starting materials. Samples of 60 vol% MA TiB_2 exhibited rates as low as 0.50 mm^3/kg.

Figure 1.3.1 shows the microstructure of the composite containing HfB_2 addition. Grains of each phase are clearly visible (the figure includes atypically large grains for clarity and composition measurements) as well as a matrix phase that does not show Z-contrast (composition) of either of the three phases. EDS measurements indicate that this inter-phase is a mixture of HfB_2 and TiB_2, which may be single phase due to the shorter diffusion distances of the smaller particles produced by milling. The ZrB_2 addition composite is not shown due to less obvious features, a consequence of the lower differences in Z-contrast.

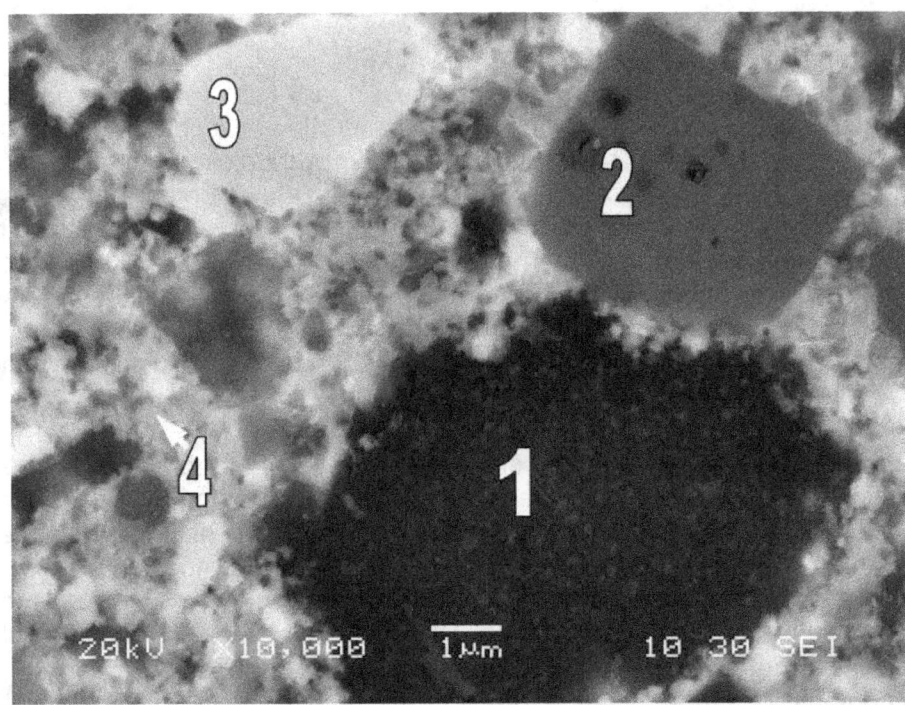

Figure 1.3.1: $AlMgB_{14}$-TiB_2-HfB_2 composite (40:30:30 vol fraction).
Areas marked 1: $AlMgB_{14}$, 2: TiB_2, 3: HfB_2, and 4: $(Ti,Hf)B_2$.

Since erosion rates at 1400 °C (leading to higher porosity) and 1500 °C (leading to larger grains) are nearly identical in the CrB_2 samples, it is likely that an intermediate consolidation temperature would result in an improved erosion rate. In the HfB_2 samples, erosion actually increased with higher sintering temperature. Interdiffusion between the diboride phases appears minimal at 1400 °C, and all composites contained three phases. Since pre-reacted powders of each pure diboride were employed in these samples, these results indicate that diffusion is sluggish at the lower sintering temperatures. If diffusion rates were sufficiently high for complete mixing of the phases, grain growth would also be unacceptably high. At 1500°C, both the Cr- and Zr-boride additions remained nominally three phase, while the Hf boride addition showed significant evidence of interdiffusion (Figure 1.3.2).

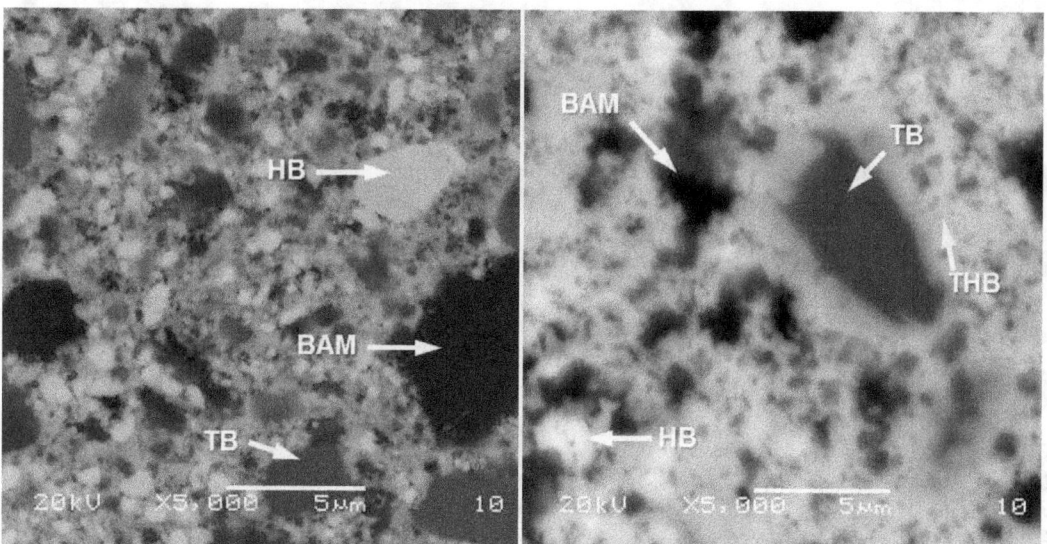

Figure 1.3.2: Backscattered images of $AlMgB_{14}$-TiB_2-HfB_2 composite pressed at 1400° (left) and 1500°C (right). $AlMgB_{14}$ (BAM), TiB_2 (TB), HfB_2 (HB), and mixed $(Ti,Hf)B_2$ (THB) present. Note diffusion gradient around large TiB_2 grain in 1500 °C sample.

Significant interdiffusion between the Ti and Hf phases was observed to occur after pressing at 1500°C, forming a matrix of almost entirely single-phase $(Ti,Hf)B_2$. This process also appears to have eliminated the majority of the finest grains in the composite, resulting in higher average grain size. Despite the increase in density of ~2%, the erosion resistance decreased as a result of pressing at 1500 °C. This fact alone cannot conclusively establish whether this is solely a function of grain size, or if the transition from a three- to two-phase composite resulted in an increase in the erosion rate.

While the CrB_2 addition resulted in an increase in density, it had little effect on erosion resistance as the increase in density was accompanied by grain growth (Figure 1.3.3). Since the hot pressing temperatures were selected based on prior experience with $AlMgB_{14}$-TiB_2 composites, fine-tuning of the temperature may result in an improvement in erosion rate, where there is a balance between densification and grain growth. Compositions containing ZrB_2 showed the largest increase in density corresponding to an increase in sintering temperature and also exhibited the highest erosion resistance, this implies that 1500°C may be closer to the optimum sintering temperature for ZrB_2 additions, while farther off with respect to the CrB_2 and HfB_2 additions.

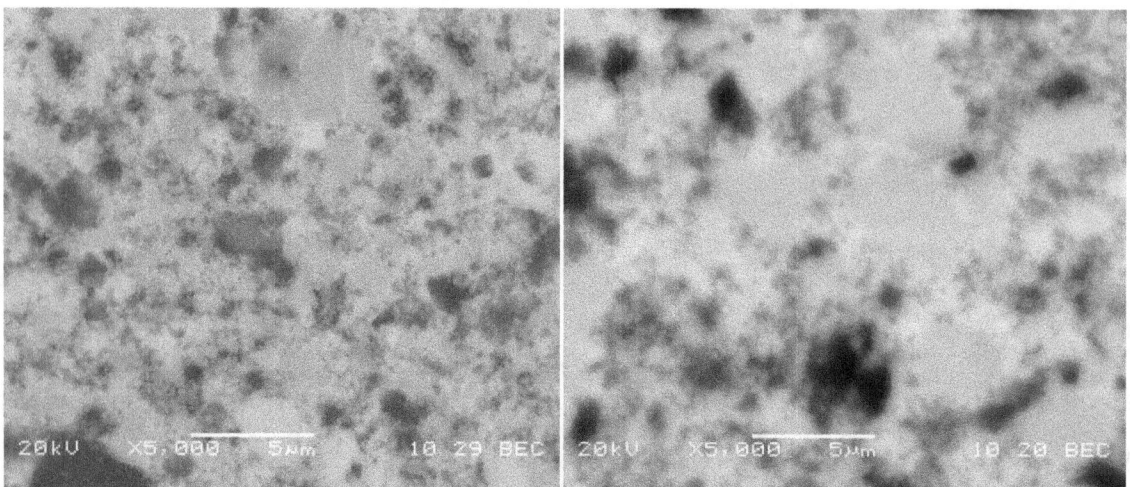

Figure 1.3.3: SEM images of AlMgB$_{14}$-TiB$_2$-CrB$_2$ composite pressed at 1400°C (left) and 1500°C (right). AlMgB$_{14}$ (black), TiB$_2$ (grey), and CrB$_2$ (slightly brighter grey) are clearly seen.

The ZrB$_2$ sample pressed at 1500 °C showed a significant increase in density over 1400°C and exhibited erosion resistance comparable to similar AlMgB$_{14}$-TiB$_2$ composites. Even though the sample experienced a low erosion rate, inspection of the erosion crater revealed multiple "pockets" of localized erosion. These "pits" within the typical crater were investigated for differences in microstructure from the surrounding regions. In addition to the three phases present in the majority of the composite, the pits were found to contain a fourth phase formed by the mixture of the finest of the TiB$_2$ and ZrB$_2$ grains. While it is not clear why there was significant diffusion in these localized regions, it indicates that the mixed diboride phases are inferior to the three-phase composite. The three-phase composite likely offers more strengthening mechanisms in the way of crack deflection between a higher fraction of phase boundaries.

Both HfB$_2$ and ZrB$_2$ are known to be completely soluble in the TiB$_2$ structure at high temperatures and are also know to form solid solutions during MA. If MA diboride powder was prepared from elemental Ti, (Zr or Hf), and B, it should produce a two-phase composite similar to the AlMgB$_{14}$-TiB$_2$ composites. This was attempted with MA Ti$_{.5}$Zr$_{.5}$B$_2$ powder added to AlMgB$_{14}$. The resulting microstructure is shown in Figure 1.3.4. Comparison of the additions shows a three-phase composite when using MM powder and a two-phase composite retaining the solid solution of the MA powder.

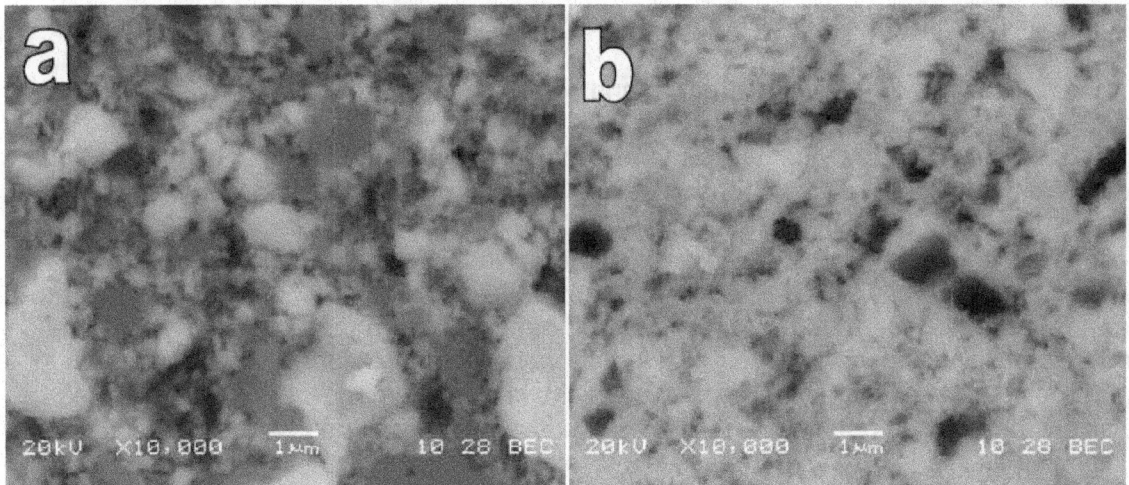

Figure 1.3.4: (a) MM AlMgB$_{14}$-TiB$_2$-ZrB$_2$ and (b) MA AlMgB$_{14}$-(Ti,Zr)B$_2$. Black (BAM), white (ZrB$_2$ on left, (Ti,Zr)B$_2$ on right), and grey (TiB$_2$) phases present.

A further subset of the group of compositions involving secondary reinforcement phases include those containing CrB$_2$, which provides an added advantage of increased oxidation resistance, as shown in our previous work. The microstructure of an AlMgB$_{14}$ – TiB$_2$ - CrB$_2$ sample is shown in Figure 1.3.5a. The low Z-contrast frustrates determination of the presence of solid-solubility, if any. EDS indicates that there may be some diffusion of Ti into CrB$_2$ and vice-versa. Figure 1.3.5b shows a composite produced by MA of Al, Cr, and B. The figure clearly illustrates the fine and uniform microstructures that can be obtained by the simultaneous MA of insoluble phases. Erosion resistance is not expected to be exceptional as the matrix phase is AlB$_{12}$. This microstructure is similar to the Al, Mg, Ti, and B in-situ composites described above. This sample could be reproduced with different starting powder mixtures to produce an AlMgB$_{14}$-CrB$_2$ composite.

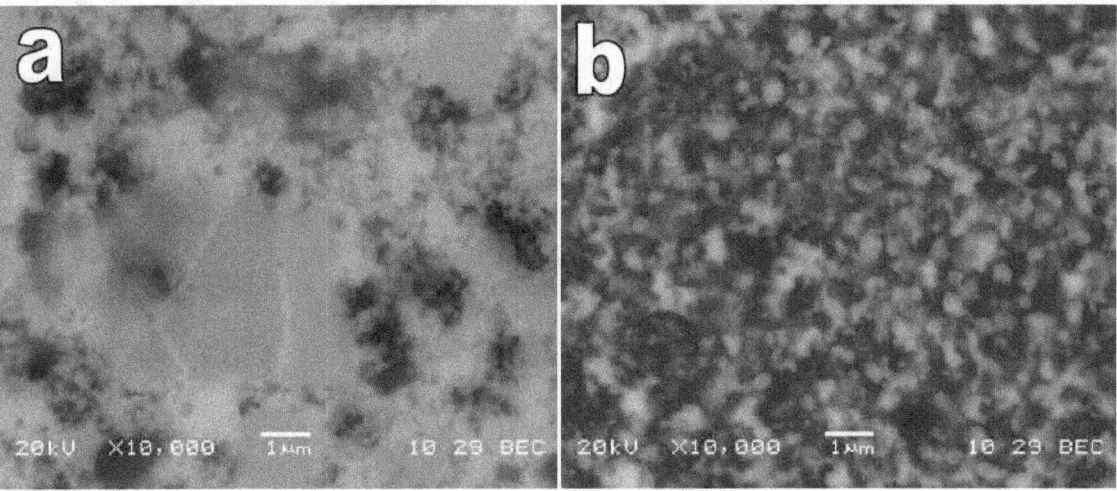

Figure 1.3.5: (a) MM AlMgB$_{14}$-TiB$_2$-CrB$_2$ and (b) MA AlB$_{12}$-CrB$_2$.

Depending on their microstructures, three-phase composites could offer superiority to the typical two-phase composites, for instance, three phases may mutually impede grain growth to an extent beyond that of only two phases and simultaneously provide more opportunities for crack deflection and energy dissipation. The alternate microstructure is a two-phase composite, wherein the two diboride phases diffuse into one. This is possible because many diborides are partially if not fully soluble in each other at elevated temperatures; this could occur if there is sufficient time for diffusion [Wang03,Zdaniewski87,Millet96]. This arrangement may offer the benefit of both constituents (i.e. high strength and corrosion resistance).

One pathway to synthesis of these mixed di-borides with minimal grain growth is the MA of the constituent elements into a nearly amorphous, single-phase powder. This is known to be possible for many diboride systems, especially $(Ti,Zr)B_2$ [Moriyama98,PetersWOM09]. Though not discussed above, this is especially interesting because this mixture is metastable due to the high mechanical deformation. While many diborides have partial or complete solubility at elevated temperatures, i.e. >2000 ºC, little research has been performed on the lower temperature stability of these mixtures. If low temperature mutual solubility is low, these metastable powder mixtures may decompose into two immiscible nanoscale phases. Here, particle size is determined not by milling, but by diffusion, which can be controlled through heating rates and sintering temperatures. As indicated by the previous series of Cr-, Zr-, and Hf-boride experiments, interdiffusion begins to occur at the higher sintering temperatures (i.e., 1500ºC) but may not be kinetically favorable at lower temperatures.

One step further from the mixed diborides, samples with either ZrB_2 or HfB_2 were prepared by MM of the reinforcement boride with $AlMgB_{14}$ powder, without any TiB_2 fraction. Mixing appears qualitatively similar to that obtained in samples containing TiB_2 as shown in Figure 1.3.6. TiB_2 is similar in many respects to ZrB_2 and HfB_2, due largely to the similarities of the metal atoms that comprise group 4 of the periodic table. As the $AlMgB_{14}$ and TiB_2 composite's exceptional wear properties are due to the excellent interfacial bonding, ZrB_2 and HfB_2 can be expected to behave similarly [Cook10]. Slight changes in surface energy and interfacial properties may offer improvements in wear resistance as well. From the applications perspective, there are environments in which Zr- or HfB_2 may offer distinct advantages, for instance in corrosive or oxidative environments [Mroz93,Gasch04,Weng09].

Because of the significant amount of prior data collected for the 60 vol % TiB_2 composition, identical samples with 60 vol% of either ZrB_2 or HfB_2 (ZB60 and HB60, respectively) were prepared. Figure 1-5 shows typical microstructures of each of these compositions; while the microstructures are generally quite similar, there are a few subtle differences. While the overall size distribution of the three samples is similar, there appears to be a "cut-off" with few grains under ~100 nm in the ZB60 and HB60 samples. This "cut-off" may indicate coarsening, which further implies higher diffusion rates in these materials. The morphology of each is also slightly different. In the TB60 composite, most of the TiB_2 grains are dispersed in a thin $AlMgB_{14}$ matrix. In the ZB60 and HB60 samples, there exists more bridging between the diboride grains, forming a network or skeleton. Additionally, there appears to be a higher degree of faceting within the ZB60 and HB60 composites, which may result from higher surface energies.

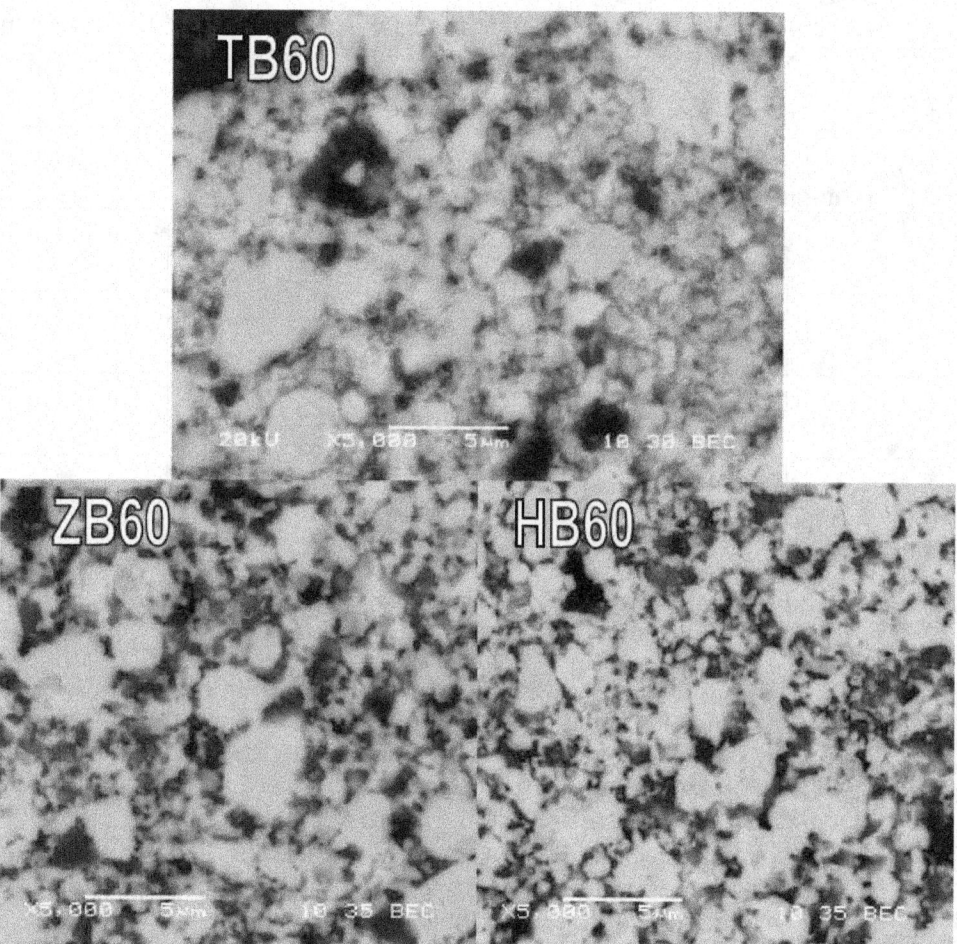

Figure 1.3.6: Typical microstructure of TB60, ZB60, and HB60 samples. Diborides represented by the brighter phase in each micrograph. (Note the similarity in microstructure between the two samples, indicating that the mixing process developed for TiB_2 applies to other Group IV borides as well.)

Preliminary investigation of the interfacial bonding between phases by introduction of Vicker's indentation cracks has given results comparable to those observed in the $AlMgB_{14} - TiB_2$ composites, indicating the interfacial bonding may also be strong in these new composites. Erosion testing of the composites containing ZrB_2 and HfB_2 was compared with $AlMgB_{14} - TiB_2$ compositions, as well as with WC. Evaluation of these samples was based on more aggressive testing protocols than earlier samples in order to establish the performance limitations of these composites.

The measured density and steady-state ASTM erosion rate for these samples are listed in Table 1.3.2, along with that of RocTec 500 and K68 WC. It is noteworthy that the HB60 sample exhibits nearly the same erosion behavior as the TB60 sample. Taking into consideration that these are the first samples synthesized with these compositions, refinements in processing and microstructure may offer significant improvement; HfB_2 additions have the potential to surpass the wear resistance offered by TiB_2.

The wear rate of the zirconium-based sample was roughly twice that of the titanium- and hafnium-based composites. The presence of agglomerates in this sample likely contributed to its poorer performance. These agglomerates were not examined in detail, although preliminary inspection suggests that there may have been some reactivity between ZrB_2 and Fe impurities. Again, this was a preliminary study and these two new additions definitely show potential for further study.

Table 1.3.2: ASTM dry erosion rates of TB60, ZB60, and HB60 samples compared with commercial WC materials.

Material	Density	Erosion Rate
	%	mm^3/kg
TB60	97.7	0.20
ZB60	97.3	0.53
HB60	98.0	0.28
Roctec 500 WC	~100	0.15
K68 WC	~100	14.32

1.3.2: SiC Whisker Reinforcement

Three different toughening agents were added to the composite in an attempt to increase fracture toughness. These were:

(1)   a highly-ductile Co-Mn alloy - Developed early in this program as a next-generation binder/sintering aid superior to pure Co, this Co-Mn alloy [U.S patent 6,921,422] is stronger and more ductile than pure Co.

(2)   SiC whiskers (SiCw) - High-aspect-ratio whiskers of various compounds are sometimes added to ceramic materials to add toughness by crack deflection, whisker bridging, and whisker pullout. For example, the toughness of alumina can be increased from 3 to 8.5 MPa$\sqrt{m}$ by the addition of 20 volume percent SiC whiskers [Becher90].

(3) Si_3N_4: The presence of SiAlON has been shown to raise toughness in some other ceramic materials. SiAlON can be induced to form in $AlMgB_{14}$-TiB_2 composites by reacting the Al_2MgO_4 (spinel) impurity phase already present in the composites with Si_3N_4.

Greenleaf Corporation collaborated with Ames Laboratory on development of bulk $AlMgB_{14}$-TiB_2 machining tools with improved fracture toughness. A small quantity of SiC whiskers (SiC:w) was transferred from Greenleaf to Ames for processing studies designed to produce a uniform composite. SiC whiskers are currently used as reinforcement for ceramic cutting tools (Al_2O_3, TiB_2, etc.) [Deng98,Deng05]. The addition of SiC:w in ceramic-based tools results in a marked increase in fracture toughness. For example, addition of 10 to 30 vol. % SiC:w in TiB_2 resulted in a 40% increase in fracture toughness [Kamiya95]. In another example, the fracture toughness of alumina was increased from ~3.0 MPa$\sqrt{m}$ to 8.5 MPa$\sqrt{m}$ with the addition of 20 vol. % whiskers [http://www.ms.ornl.gov/researchgroups/process/cpg/sic.htm]. A limited study was planned wherein 5 and 10 vol.% SiC:w was added to $AlMgB_{14}$.

The compatibility of SiC with TiB_2 suggests the possibility of its use in the $AlMgB_{14}$ composites. The first step is testing the stability of SiC with $AlMgB_{14}$. A preliminary sample was prepared in which baseline powder was media-less milled with 5 vol% SiC whiskers, and subsequently hot pressed at 1500°C. Media-less milling was performed in order to prevent fracture of the high aspect ratio whiskers, and to provide uniform mixing. SEM of the resulting powders (Figure

1.3.7) reveals a homogenous mixing of the constituents. On the other hand, a fracture surface of the consolidated sample (Figure 1.3.8) shows many flattened agglomerates of SiC that apparently did not mix with the bulk of the sample. Close inspection of the AlMgB$_{14}$-SiC interface does show some bonding and phase stability (Figure 1.3.9). Within the AlMgB$_{14}$ matrix, few SiC particles were identified. It was not immediately clear whether that indicated the majority of whiskers were contained in the agglomerates, if the SiC whiskers were reacted/dissolved by the matrix, or if the small diameter of the whiskers makes them difficult to detect. Higher volume fractions and improved mixing methods were attempted to resolve this question.

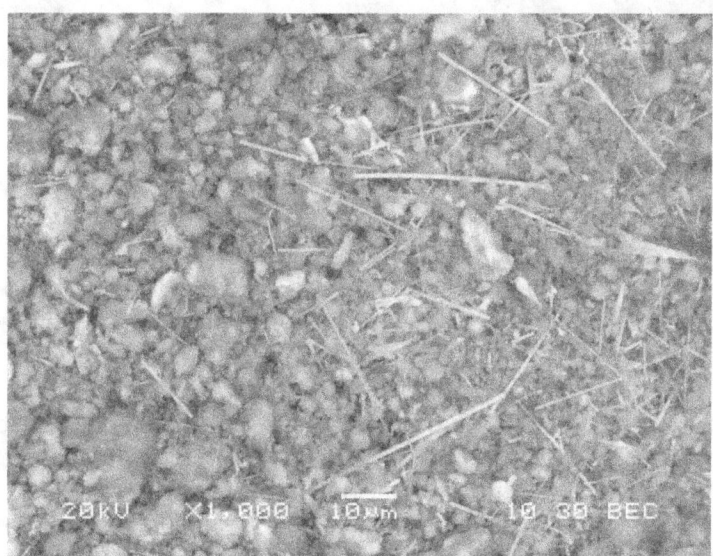

Figure 1.3.7. TB00 and SiC whisker powder after dry milling.

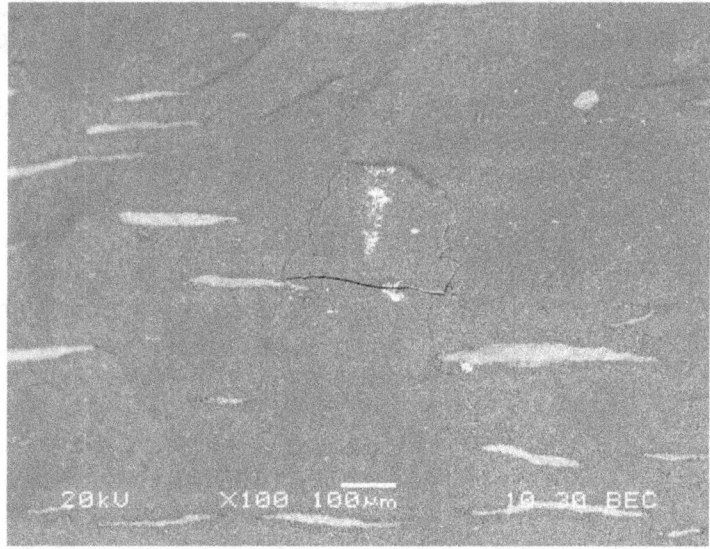

Figure 1.3.8. SiC (bright) agglomerates flattened by compaction during hot pressing in a TB00 matrix sample.

Figure 1.3.9. Boundary between TB00 matrix and SiC whisker agglomerate. Labeled whiskers show apparent bonding with the matrix, while there is little sintering of the whiskers alone (image bottom).

Due to the large aspect ratio of the whiskers, high-energy milling is not a viable mixing approach as the whiskers quickly break into sub-micron particles. In the initial attempt to blend SiC:w into the mixed-phase powders, it was found that mixing without media was insufficient to break up the SiC:w agglomerates. However, the use of low-density plastic media offers a potential means of blending the constituents and breaking up the SiC:w agglomerates without introducing excessive momentum transfer during the process. Figure 1.3.10 shows the result of 1 hour of mixing of SiC:w with elemental Al, Mg, and B using plastic media. Use of the elemental constituents is an attractive starting point because of the lower cost associated with industrial processing. It is seen that the whiskers are well-mixed with the $AlMgB_{14}$-TiB_2 powder, while retaining some of their high-aspect ratio characteristics. Fragmentation of the whiskers was observed, although most remained at least 10 μm in length. Additional mixing methods such as slurry processing and sonication were not available.

The polymer-milled powders were hot pressed at 1400 and 1500 °C. Figure 1.3.11 is a micrograph of a fracture surface of the 1500 °C sample, showing that the SiC phase had partially reacted with the elemental components. The morphology of the Si-rich phases does not correspond to that of the whiskers, and none of the pores present are likely due to pull-out as no exposed whiskers are seen. It is likely that the SiC:w will need to be added to pre-reacted $AlMgB_{14}$ instead of to the elemental constituents. If 1400°C is not a sufficiently low temperature to prevent reaction, Dynaforging or SPS could provide an attractive alternative consolidation route for the elemental powder mixtures. As described under Task 2, SPS may be particularly well-suited to densification of SiC whisker composites as this process appears to require lower temperature and shorter time to achieve good densification (see Task 2: Development of Scale-up Technology).

Figure 1.3.10. SiC whiskers in Z-TB00 milled with polymer media (left). As received whiskers (right).

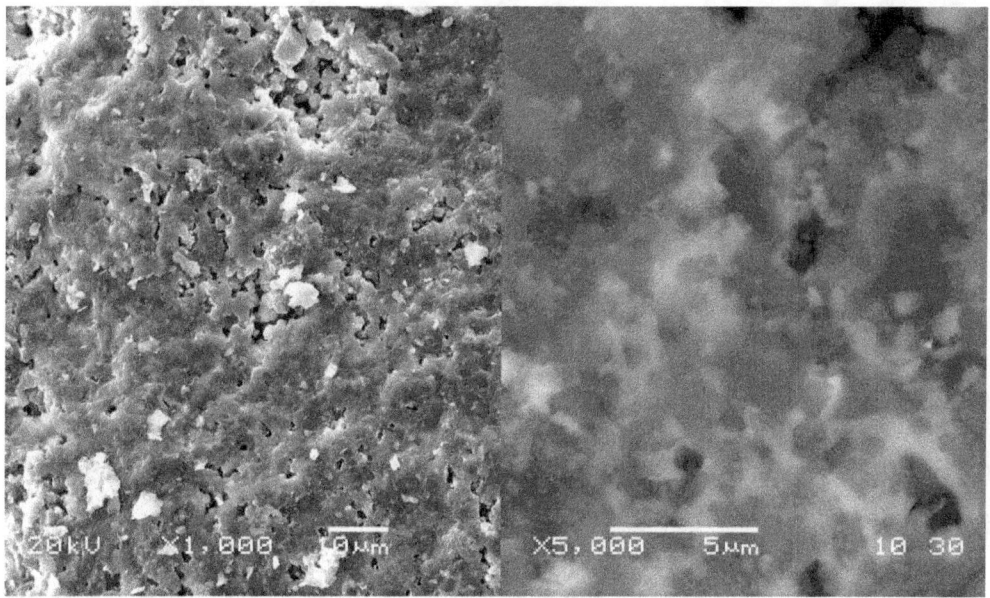

Figure 1.3.11 Fracture surface of Z-TB00 sample with 10 vol% SiC-whiskers, hot pressed at 1500°C. Topography (left) shows pores but no evidence of whisker pull-out. Composition (right) reveals Si-rich phases (light grey). Note: different magnification.

Sintering at 1500°C resulted in partial dissolution of the SiCw phase; the subsequent results of hot pressing at 1400°C are shown in Figure 1.3.12a. Also shown in the right of the image is an identical sample pressed at ORNL by SPS at 1300°C. Both microstructures are similar; there is little reactivity between the phases yet it is readily apparent that SiC:w impedes densification. Scattered regions of fully-dense $AlMgB_{14}$ appear to contain little SiC:w reinforcement. As shown in Figure 1.3.12, SPS can achieve nearly identical results to conventional hot pressing with lower temperature and shorter time. Additionally, SPS can achieve higher heating rates; 50°C/min vs 20°C/min for HP. Higher temperatures coupled with high heating rate and shorter time at temperature may achieve full densification with minimal reactivity (i.e. 1400°C, 50°C/min, 10 min at temperature). (See Task 2 for further evaluation of the SPS process)

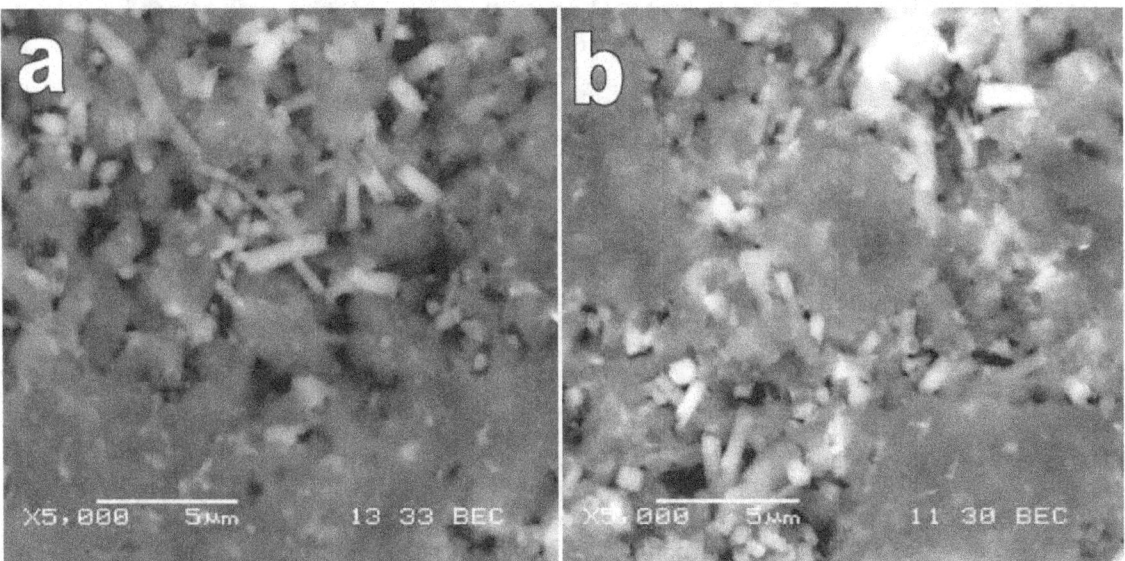

Figure 1.3.12: TB00 + SiCw samples hot pressed at 1400°C for 60 min (a) and consolidated by SPS at 1300°C for 30 min (b).

Use of pre-reacted (MM) AlMgB$_{14}$ did little to ameliorate the reactivity between SiCw and the matrix. Samples pressed at 1400 and 1500°C were similar to those prepared with elemental MA AlMgB$_{14}$. As shown in Figure 1.3.13, use of pre-reacted AlMgB$_{14}$ resulted in slightly higher porosity due to reduced sinterability and higher Fe and O impurities as a result of increased processing and handling.

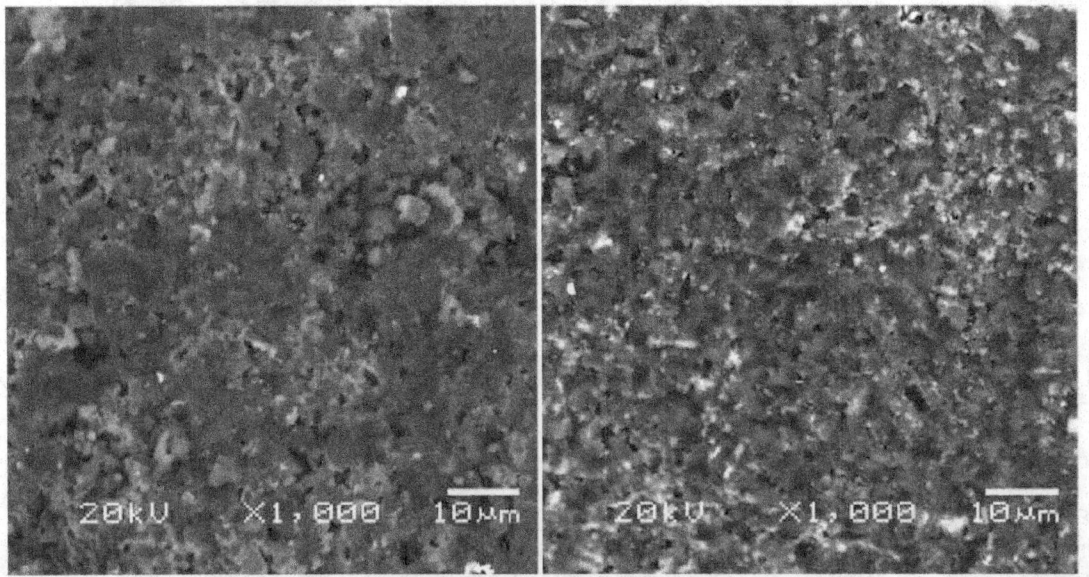

Figure 1.3.13: TB00 + SiCw composites HP at 1500°C prepared with as-milled AlMgB$_{14}$ (left) and pre-reacted AlMgB$_{14}$ (right). Bright phases are Si-rich spinel.

In order to understand the interaction between the SiC:w and each of the two primary boride phases in the composite, diffusion couples of TB00 and TB100 with SiCw were prepared. After hot pressing at 1500°C for 60 min, the left of Figure 1.6 shows an apparent reaction layer of about 3 μm and the depth at which the whiskers appear to be consumed is about 1 μm; however, these values exceed the thickness of any single whisker (~300 nm). Thus, at 1500°C, the reaction is excessive, and shorter times at high temperatures will be required to produce densification without complete reaction of the whiskers. A similar diffusion couple of TiB_2 with SiCw was also prepared, shown in the right of Figure 1.3.14, where no reaction layer is visible, yet the contact between the two phases appears good. This is in agreement with other research, where SiC is used in a composite with TiB_2 [Deng98,Deng05,Holleck87,Torizuka95,Kamiya95]. One could employ TiB_2 as an interfacial buffer between $AlMgB_{14}$ and SiCw, preventing the reaction. $AlMgB_{14}$ is an effective sintering aid for TiB_2, reducing sintering temperatures from 1700-2000°C to 1400-1500°C. Samples are currently being prepared with 70 vol % TiB_2, 10 vol % $AlMgB_{14}$, and 20 vol % SiC whiskers. In these samples, it is expected that even if there is a reaction between the whiskers and the TB00 phase, a reasonable volume of whiskers will remain after consolidation.

Figure 1.3.14: SiCw (top) couple with TB00 (left bottom) and TiB_2 (right bottom). TB00-SiC reaction layer marked. (Reaction couple was formed at 1500°C)

Of note are the large pores in Figure 1.3.15 which may be a sign of CO gas generation based on the elements present. This is an indication that C or SiC additions may yet be effective oxide reducers. Depending on the results of this experiment and those described under Task 1.4, a mixture of SiC and Si_3N_4 could hypothetically be able to remove oxides and form SiAlON simultaneously.

Since even a small amount of $AlMgB_{14}$ can be an effective TiB_2 sintering aid, as low as 15 vol% at 1400°C with current techniques, it is likely that TiB_2 with moderate additions of both SiC and $AlMgB_{14}$ can be processed into dense, tough ceramics at relatively low temperatures, where the low fraction of $AlMgB_{14}$ minimizes any unwanted reactions. Dry mixing and milling were also used to mix whiskers with boride constituents, but these did not completely eliminate whisker agglomerates. Ultrasonic mixing in ethanol does break up the agglomerates, but the components (each of which possesses a different density) separate out of the liquid at different rates.

Incorporation of SiC whiskers in the boride composites seems to show that the SiC phase may getter some oxygen from the material. Certain regions (shown in Figure 1.3.15) were found to be porous, with no visible whiskers. Generally, the porous regions in such samples tend to be rich in whiskers, which inhibit densification. In the regions shown in the figure, EDS reveals the presence of Si, but no C. This suggests that upon reaction of SiC with TB00, the C may have been converted to a gas phase. The likely phase would be CO (due to the high temperature) which could also explain the porosity. Consequently, to form CO the carbon must be reacting with oxides, the majority of which are tied up in the spinel phase. Because this result was only found in certain regions, there appear te be certain required conditions for the reaction to proceed. The exact nature of these conditions is unknown. Additionally, if C from SiC is capable of reducing oxides, there is no thermodynamic reason that pure C additions could not do the same although this effect was not found in past samples with graphite additions. Vacuum hot pressing may also aid in densification by removing the gas phase as it is formed. It should be noted that the ideal case for spinel removal during sintering would be CO formation as there would be no residual oxide phase.

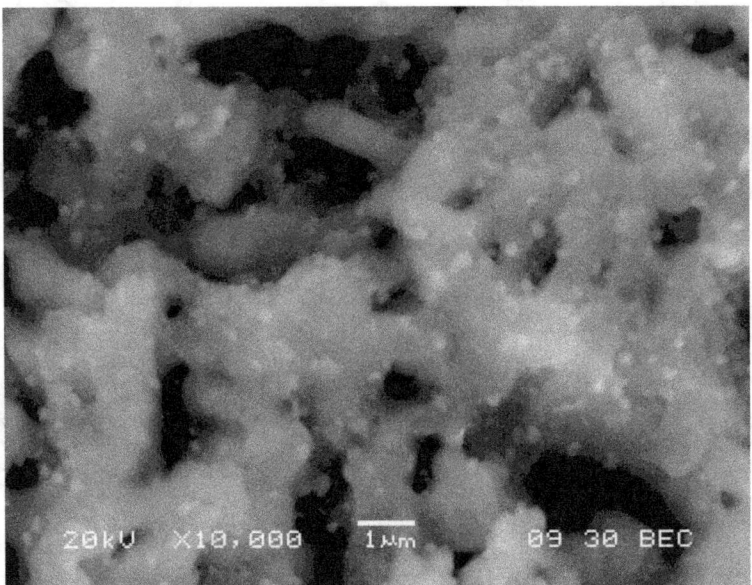

Figure 1.3.15: TB00 + SiCw sample pressed at 1400°C for 60 min. Light spots are Si-rich particles. No C was present in such regions.

It could also be interesting to synthesize a sample of $AlMgB_{14}$ with TiB_2 in whisker form. TiB_2 whiskers can be synthesized by combining TiO_2 and B_2O_3 with carbon according to the following reaction:

$$TiO_2 + B_2O_3 + 5C \rightarrow TiB_2{:}w + 5CO(\uparrow)$$

One possible issue might be that the exceptionally strong bonding between boride phases might hinder the whisker pull-out mechanism for toughening.

Achieving a homogeneous distribution and avoiding reaction with the constituent material are also issues associated with SiC whisker additions. Improved milling techniques were developed and these have produced powders exhibiting a significant improvement in the distribution of whiskers compared with initial attempts. SiC:w requires low-energy milling in order to prevent damage and excessive fracture of the whiskers. Yet comminution of TiB_2 and $AlMgB_{14}$ agglomerates require higher-energy milling. To this end, an intermediate process was developed in which polymer media were used to mix the three components. Additional powder has been prepared where milling of the TiB_2 and $AlMgB_{14}$ was performed with steel media, then the SiC whiskers were added and the steel media replaced with polymer media for additional mixing. These improved powder samples may be promising for consolidation by SPS. In addition, higher TiB_2 concentrations are expected to mitigate some of the SiC-$AlMgB_{14}$ reaction issues. A beneficial side-effect of the reactivity issues of $AlMgB_{14}$ and SiC has been evidence that the carbon in the SiC phase may be acting as an oxygen getter. Using both SiC whisker and SiC powder, milled and pressed samples have shown indications of spinel reduction by XRD.

1.3.3: Co-Mn Binder

Advanced materials for superior wear resistance must possess both good hardness and fracture toughness, where hardness is loosely defined as resistance to plastic indentation (shear) and toughness is a measure of a material's ability to absorb an impact without catastrophic fracture. The classic example is cemented carbide; WC cermets with Co additions in the range of 3 to 13 weight percent are employed for machining cast iron, non-ferrous metals, and non-metallic materials. Grades containing up to 25 weight percent binder are normally used for non-machining, high impact applications. Hardness decreases with increasing Co content and with increasing WC particle size. Cemented carbides are prepared by liquid phase sintering with a liquid phase eutectic of C and Co which is formed at 1320°C at a composition of 11.6 at. % C.

Most industrially useful wear-resistant materials contain some amount of a ductile binder phase to increase toughness and also to improve sinterability during processing. While the performance metric for this project is based on the material's abrasive and erosive wear characteristics, an ancillary objective is to achieve the unprecedented combination of a hardness greater than 30 to 35 GPa with a toughness of at least 10 MPa√m. Previous work demonstrated that these values are achievable if the hard boride is uniformly mixed with a binder and rapidly consolidated under high pressure to full density. Task 1.2 involves examining various methods of blending the boride composite with a highly ductile binder, Co-Mn, and optimizing the consolidation process to generate ultra-hard and tough materials. This task is related to Task 2 (Industrial-Scale Processing) in that results from the lab-scale samples will be carried over to the synthesis efforts of Oak Ridge and the consolidation efforts of Carpenter.

A successful binder must exist as a liquid phase within a temperature range that avoids unwanted decomposition of the active material, while also possessing a similar (or lower) surface energy to enable good "wetting" of each grain. Furthermore, the binder must possess sufficient ductility (as measured by percent elongation at fracture) to absorb and dissipate the energy associated with an advancing crack tip, while retaining adequate strength to prevent failure under typical tensile, torsional, or shear loading. Several requirements exist for liquid-phase sintering. First, the temperature must be sufficiently high so that the binder phase becomes completely liquid. A favorable contact angle must exist between the liquid binder phase and the solid base material. In other words, the relative surface energies of the two phases must be sufficiently low so that the liquid "wets", or completely covers each hard particle. Moreover, an appropriate volume fraction of binder phase must be present. In the case of insufficient volume fraction of binder, the composite may contain excessive porosity and lack mechanical strength. In the case of excessive amounts of binder phase, the mechanical properties of the material will converge to that of the binder itself, rather than that of the harder base material. In addition, excessive binder can result in liquid phase "squeeze-out" during sintering and undesired shape changes. Figure 1.3.16 shows the microstructure of a sample of $AlMgB_{14}$ – 70 wt. % TiB_2 that was sintered with 15 wt. % Co-Mn binder.

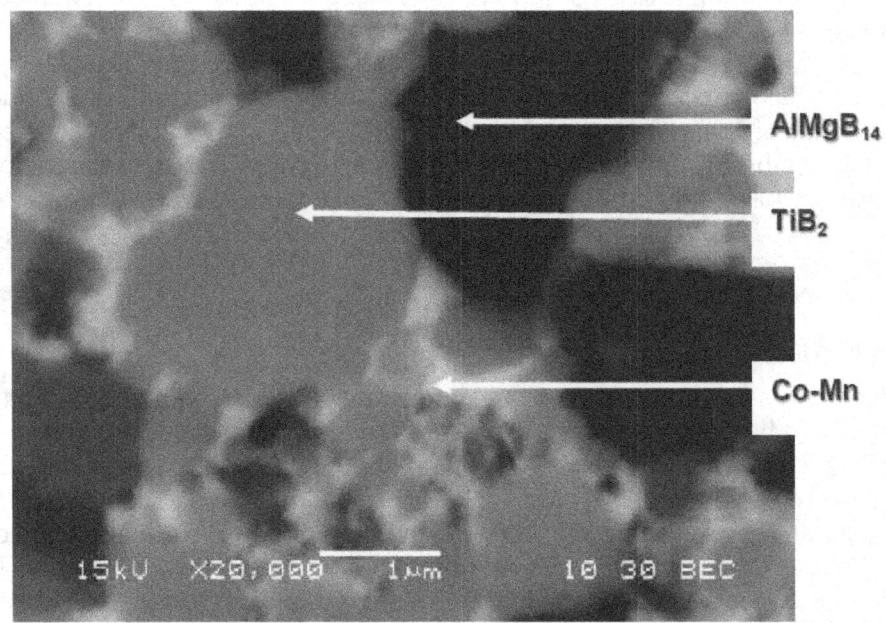

Figure 1.3.16. SEM micrograph of mixed boride composite ($AlMgB_{14}$+ TiB_2) consolidated with 15 wt. % Co-Mn binder. In the micrograph, #1 refers to the $AlMgB_{14}$ phase, #2 refers to the TiB_2 phase, and #3 refers to the Co-Mn binder.

Comparison of the Vicker's indentation crack lengths with those of binderless boride composites show that the binder increases toughness; the average crack length from a 1 kg load in the above sample was 41 microns, whereas without binder the average crack length was close to 70 microns. In addition, there was an increase in hardness due to a reduction in porosity. These results are encouraging and serve as a starting point for future work in this task. They also illustrate two important points. First, the distribution is relatively uniform with no large agglomerates or regions completely devoid of the binder. Second, the grain size of the TiB_2 and $AlMgB_{14}$ phases is too large and requires additional refinement to achieve optimal properties.

The initial exploratory effort involved a sample of hot-pressed $AlMgB_{14}$ - 70 wt% MA TiB_2, crushed by milling, and mixed with atomized powder of a Co-Mn alloy. In order to evenly coat the composite powder and prevent agglomeration of the alloy, the two loose powders were shaken in a Spex vial without media for 24 hrs, where the hard composite grit was allowed to act as its own media on the ductile alloy particles (self-comminution). An additional milling step employing ¼" media for 2 hours was applied to break up any remaining agglomerates. The mixture was then hot pressed for 10 min at 1400 °C with a minimal applied load of 35 MPa (5 ksi). The reduced pressure and shorter time compared with consolidation of binder-less composites minimizes reaction and squeeze-out of the soft binder. The resulting compact exhibited little porosity by SEM, although the microstructure appeared somewhat inhomogeneous. Vickers indentation cracking indicated excellent bonding between the phases. As seen in Figure 1.3.17, darker agglomerates or under-milled composite particles are reduced in Co-Mn binder phase (brighter contrast). There are also areas, indicated by arrows, which appear to be regions of segregated TiB$_2$ intermixed with a mixed Al-Mg-Ti-Fe oxide. The second image in the figure shows darker regions without Co-Mn which contain higher porosity.

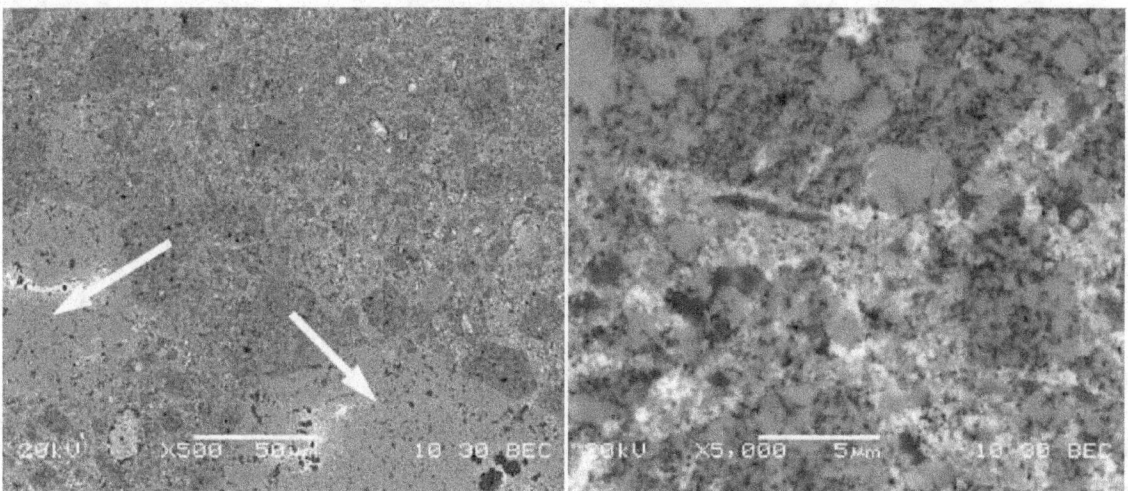

Figure 1.3.17: 70 wt% TiB$_2$ composite with 15 wt% Co-Mn. Left image shows large TiB$_2$ regions indicated by arrows and agglomerates (dark) with reduced binder content (bright). A similar agglomerate is also shown in the upper portion of the right image, with Co-Mn more evenly mixed in the remainder of the image.

An x-ray diffraction pattern of the reacted compact is shown in Figure 1.3.18, with the primary phases indexed as AlMgB$_{14}$ along with TiB$_2$ and MgAl$_2$O$_4$. The resultant alloyed compact was then ball milled to obtain a fine powder suitable for mixing with binder phase additions. Fifteen weight percent of gas atomized Co-Mn was added to the composite powder and vibratory mixed for 24 hours followed by 2 hours of ball milling with four ¼ inch steel balls. A final vibratory mixing of 2 hours completed the powder processing.

Two lots of the mixed powders were hot pressed at 1400°C under flowing Ar. Sample 031307-1 was pressed for 10 minutes and sample 031307-2 was pressed for 5 minutes. Backscattered electron images are shown in Figure 1.3.19 for 031307-1 and Figure 1.3.20 for 031307-2. The compact hot pressed for 5 minutes shown in figure 7 contained a high level of porosity. As can be seen in Figure 1.3.19a, variations in porosity exist within the sample. The density of the compact hot pressed for 10 minutes shown in Figure 1.3.19 was improved over the 5-minute sample, although some porosity was still evident.

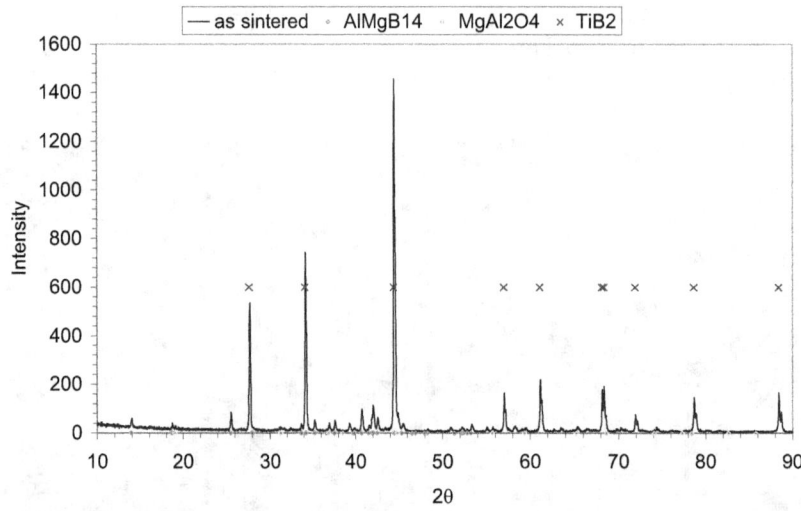

Figure 1.3.18: X-ray diffraction pattern of AlMgB$_{14}$/TiB$_2$ composite used for Co-Mn binder phase addition studies.

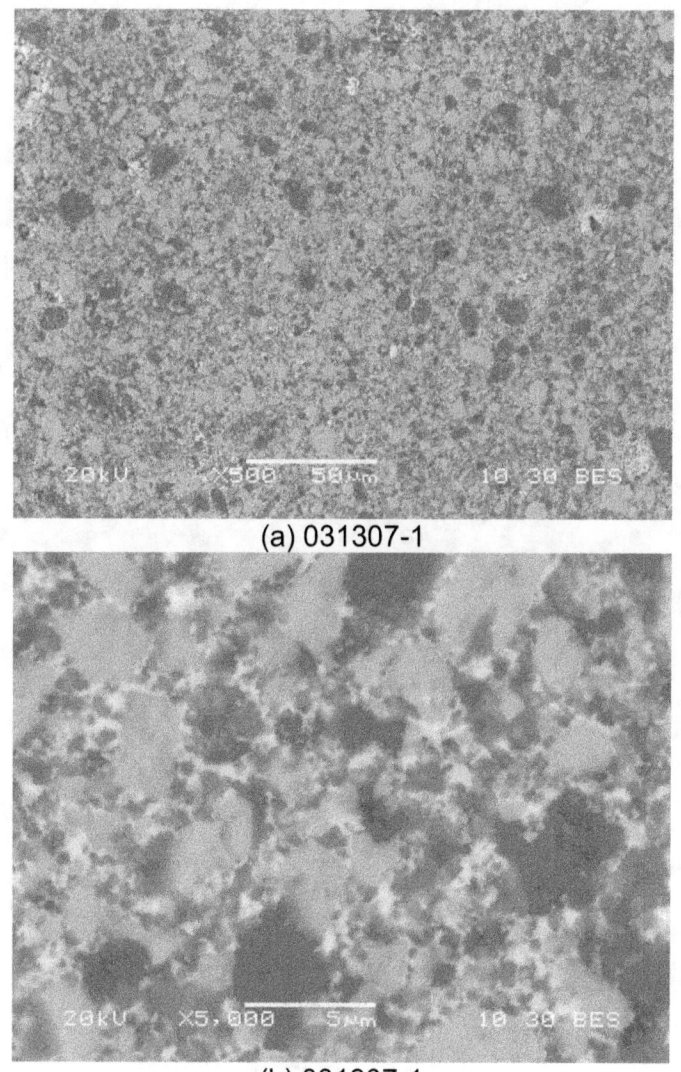

(a) 031307-1

(b) 031307-1

Figure 1.3.19. Backscattered electron images of AlMgB$_{14}$ + TiB$_2$ powders containing 15 weight percent Co-Mn hot pressed at 1400°C for 10 minutes.

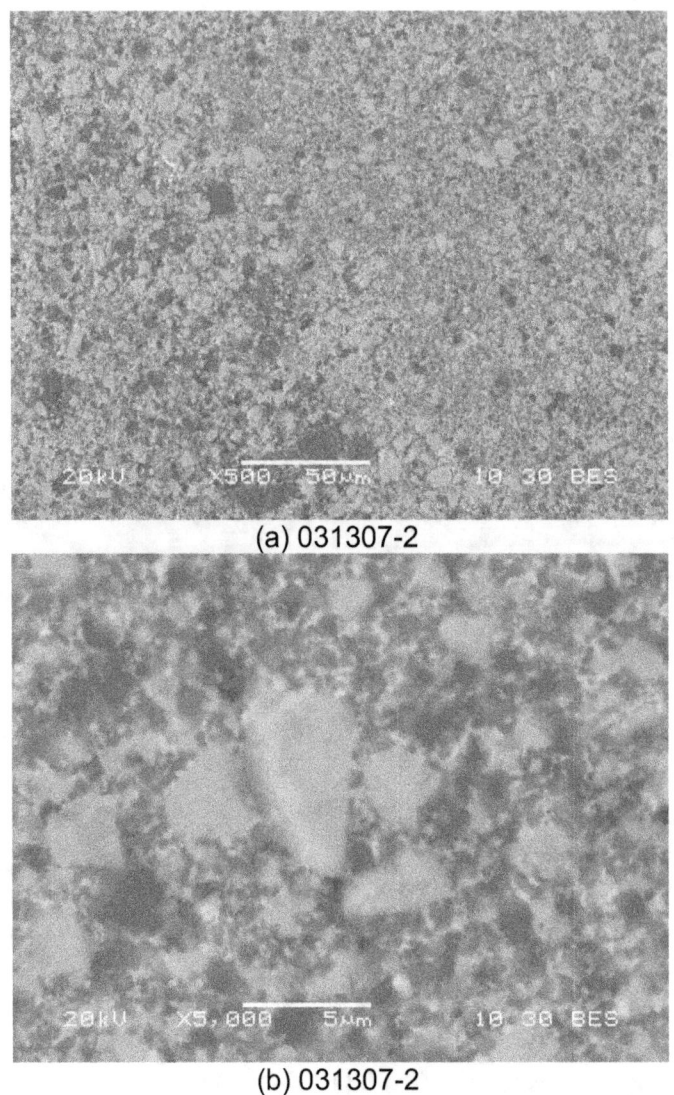

(a) 031307-2

(b) 031307-2

Figure 1.3.20. Backscattered electron images of AlMgB$_{14}$ + TiB$_2$ powders containing 15 weight percent Co-Mn hot pressed at 1400°C for 5 minutes.

The average hardness of this sample (031307-1) was 34 GPa, with an average crack length of 50 microns (2c). Micrographs of indentation cracking were analyzed for average crack length. The elastic modulus for this sample was found to equal 439 GPa from ultrasonic time-of-flight propagation studies. These results were combined through the Palmqvist equation to determine the indentation toughness, which was calculated to equal 6.2 MPa√m. This is a significant improvement over earlier samples and is an exceptional toughness for a sample possessing a hardness of 34 GPa. Further improvement might be achieved through refinement of the TiB$_2$ phase. Improved blending and reduced particle size of the AlMgB$_{14}$/TiB$_2$ composite material is expected to result in improved densification of the final hot pressed compact and an increase in toughness.

Additionally, pre-reacted AlMgB$_{14}$ + 50 wt% TiB$_2$ blended with 15 wt% Co-Mn gas atomized binder powder was consolidated at Carpenter Powder Products by the Dynaforge process for 10 seconds at 1200°C with an applied pressure of 120 ksi. This sample, designated JH040207-A, exhibited promising results for the initial attempt at rapid consolidation of a complex ceramic system. However, metallographic examination of this specimen indicated that improvements in density and binder distribution are still needed.

Two companion hot pressings of the powder batch were performed at Ames Laboratory utilizing higher temperatures. Sample JH040207-B was pressed at 1400°C for 10 minutes at 5.0 ksi while JH040207-C was pressed at 1400°C for 10 minutes at 11.6 ksi. The JH040207-B sample contained high porosity while the JH040207-C sample exhibited much higher density but also experienced squeeze-out of the Co-Mn binder.

Optical micrographs of the Dynaforged and higher pressure hot pressed samples are shown in Figure 1.3.21. The $AlMgB_{14}$ appears dark gray while the TiB_2 is blocky white. It can be seen that the Dynaforged sample has visible porosity in addition to Co-Mn inclusions. These inclusions, indicated by arrows, are nearly spherical indicating that the applied pressure was insufficient to squeeze the material into the adjoining pores. While some porosity is present in the hot pressed sample, there is no indication of large-scale binder material inclusions.

Figure 1.3.21. Optical metallography of hot consolidated $AlMgB_{14}/TiB_2$ blended with Co-Mn binder followed by second hot consolidation step. JH040207-A (top) was Dynaforged and JH040207-C (bottom) was hot pressed. Arrows indicate Co-Mn binder particles.

Figure 1.3.22 shows two additional views of the Dynaforged microstructure. The top micrograph demonstrates the distribution of the Co-Mn inclusions. It can be seen that most of these spherical inclusions possess significant void space, indicating that some melting occurred. The bottom figure is a higher magnification image of one such inclusion with what appears to be liquid flow into the boride matrix. It appears that the lower temperature Dynaforge process will require longer duration at temperature during pressure application to force the binder into matrix void space. If further improvements to porosity and binder distribution are necessary, Dynaforging at higher temperatures may be required as was demonstrated for the hot-pressed sample. The 1200°C Dynaforge temperature represented the maximum temperature of the system prior to a furnace upgrade completed just as this project was ending.

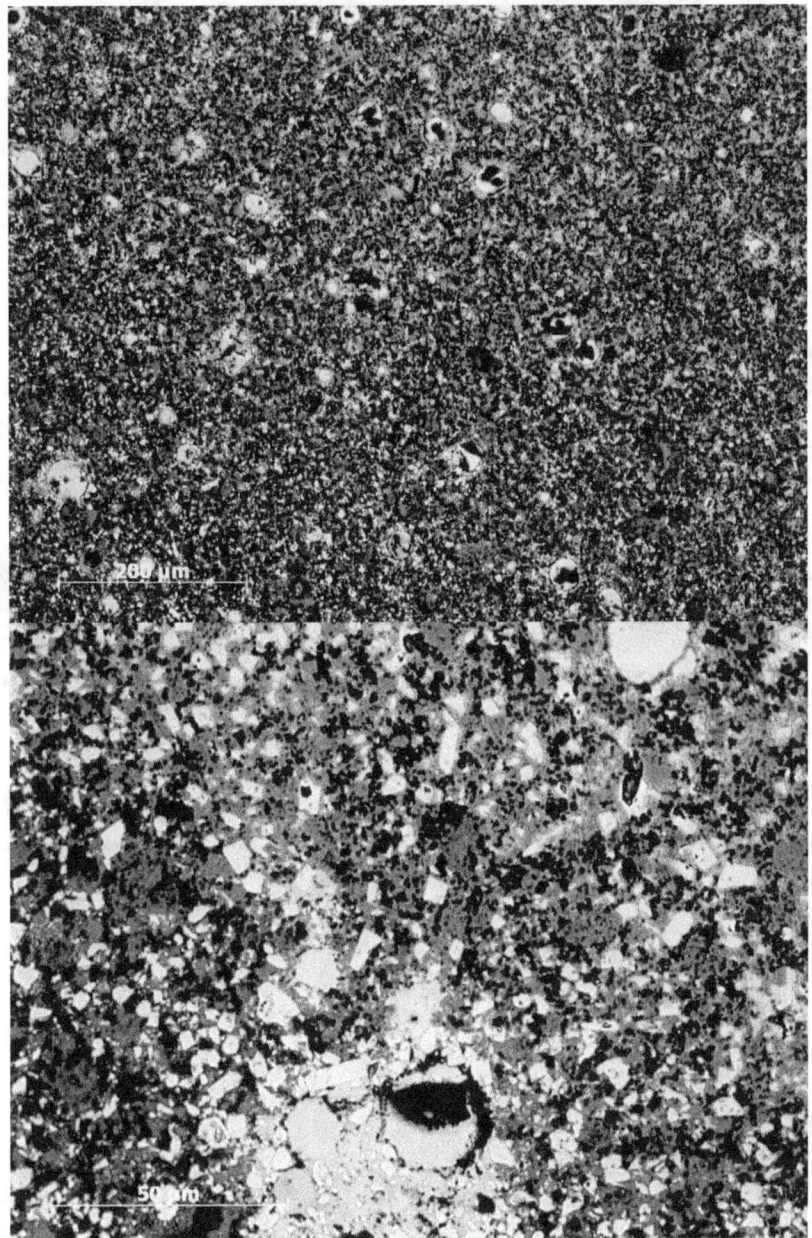

Figure 1.3.22. Dynaforged JH040207-A. Top: distribution of Co-Mn binder. Bottom: partially molten Co-Mn extending into $AlMgB_{14}/TiB_2$ matrix.

Consolidation trials of boride powder containing Co-Mn binder by Carpenter Powder Products showed encouraging results. After only a few seconds of consolidation, regions of the compact exhibit nearly full density, as shown in Figure 1.3.23. These powders were consolidated at 1200°C for 10 – 20 seconds at an applied pressure of 827 MPa. As discussed in section 1.5 below, high pressures alone are not sufficient to attain high densities at consolidation temperatures less than 1400°C. It was also shown that higher temperatures are necessary to cause flow of the Co-Mn binder, seen in Figure 1.3.23 as the bright white second phase inclusions. Optical microscopy revealed that some of the binder material had been squeezed-out during high-pressure compaction.

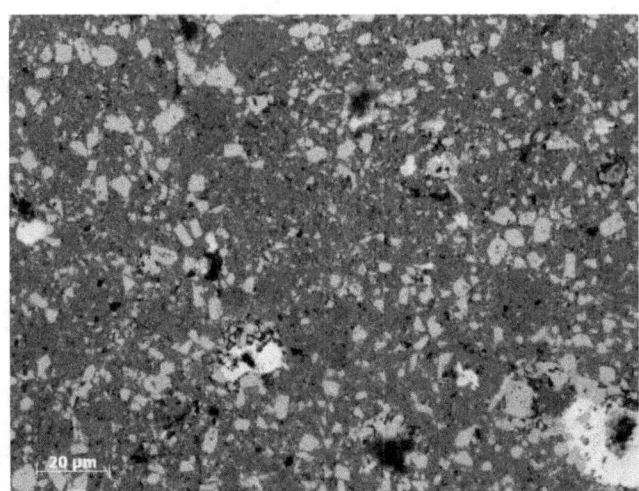
Figure 1.3.23. Dynaforged $AlMgB_{14}+TiB_2$ composite with Co-Mn binder.

Hot pressed boride compacts ($AlMgB_{14}$) comminuted by mechanical milling for 2 minutes and the resulting powder was mixed with 15 wt. % Co-Mn binder powder using a novel in-situ milling process, wherein the boride grit was employed as the active milling media. Samples from the master lot were then hot pressed at 1400°C for 10, 20, 30, or 60 min. A 1-kg Vickers indentation was applied to a polished surface of the compacts, and the lengths of the corresponding indentation cracks were measured as an indication of toughness. Results are summarized in Table 1.3.3.

Table 1.3.3. Results of Co-Mn Binder Phase Additions to $AlMgB_{14}$

	Nominal Co-Mn (wt%)	Time at 1400°C (min)	Hv (GPa)	Crack length (µm)	Toughness (MPa√m)	$K_{IC}^{1.3}H_v^{0.5}$
1	0 (baseline)	60	29	69.3	3.1	24
2	15	60	30	64.6	3.3	26
3	15	30	33	41.0	6.3	63
4	15	20	31	47.5	5.2	47
5	0 ($AlMgB_{14}$-TiB_2)	60	40	46.5	4.7	48

The ($K_{IC}^{1.3}H_v^{0.5}$) values listed in the last column of the table are of particular significance. It is this quantity that is directly proportional to the volume of material removed as a result of abrasive or erosive wear and represents an indicator of wear resistance. It is seen that the presence of the binder phase in sample #3 has more than doubled the projected wear resistance of the material by not only improving toughness, but also by improving hardness through a reduction in porosity. Extending these results to industrial-scale production and consolidation may lead to optimized densification and improve distribution of the TiB_2 reinforcement phase within the $AlMgB_{14}$ matrix.

The addition of a ductile binder phase to $AlMgB_{14}$-TiB_2 composites has been shown to improve resistance to brittle fracture, which is a desirable quality for many engineering applications. However, it is also important to maintain high hardness in order to preserve the wear resistance of the composite. In general, these properties are inversely related, and it is the goal of many research efforts to achieve the highest possible combination of the two. Cemented carbide, for example, contains varying amounts of Co as a binder, ranging from 3% to as much as 30% by weight. This provides tunable fracture toughness from 5 MPa√m at the low end, to over 25 MPa√m for carbides containing the higher percentage of binder. The Co-Mn binder used in boride consolidation is highly ductile, which is a desirable attribute for a binder phase; however, this high ductility (over 40% elongation at tensile test failure) also makes it challenging to obtain ultra-fine powder suitable for mixing. The gas-atomized Co-Mn powder falls within a size range from 10 to 100 microns, which is too large to act as an effective crack blunting mechanism without additional comminution. Development of a cost-effective process to reduce the size of gas-atomized powder and to produce a uniform cermet with the $AlMgB_{14}$-TiB_2 matrix would provide a significant benefit to commercialization of the material and to the realization of additional energy-saving applications.

Milling of $AlMgB_{14}$ (TB00) with Co-Mn and media resulted in little comminution of the Co-Mn binder; the particles were flattened and encased in $AlMgB_{14}$ agglomerates that remained in the hot-pressed or Dynaforged materials as voids or clumps of binder phase. However, when the same milling procedure was used with pure TiB_2 (TB100) and Co-Mn, no agglomerates of binder were found. It is possible that TB100 powder has a more abrasive nature than TB00 and results in significantly enhanced comminution. Powder and hot-pressed morphologies of TB00+Co-Mn and TB100+Co-Mn are contrasted in Figure 1.3.25. Co-Mn agglomerates are found in both TB00 powder and compacts, while the binder phase appears evenly distributed in both forms of TB100. (The image of TB100+Co-Mn in Figure 1.3.25 is of a fracture surface, hence the high surface roughness.) The sample showed a higher erosion rate than an identical sample pressed at 1400ºC without Co-Mn, 1.9 vs 0.75 mm^3/kg, although TB100 samples prepared with different methods have shown erosion rates as low as 0.5 mm^3/kg. The TB100+Co-Mn sample did achieve more than 98% density and was pressed at only one third of the normal pressure (35 vs 106 MPa); TB100 without Co-Mn pressed at this reduced pressure resulted in a sample of only 93% density. Ideally, the addition of Co-Mn should be able to reduce the minimum sintering temperature of TB100; however, this has not been directly demonstrated. In addition, the hardness of the TB100+Co-Mn sample was equal to that of the TB100 compacts pressed without binder; 23 GPa (3-kg Vickers). It should be noted that the loads used for measuring hardness in this study are significantly higher than those typically reported in literature (3 kg vs 1kg) and will generally translate to lower hardness values by reducing or avoiding the indentation size effect [Braz04,Chung07].

Figure 1.3.24. TiB$_2$ platelets (grey) and Co-Mn atomized spheres (bright contrast) after milling.

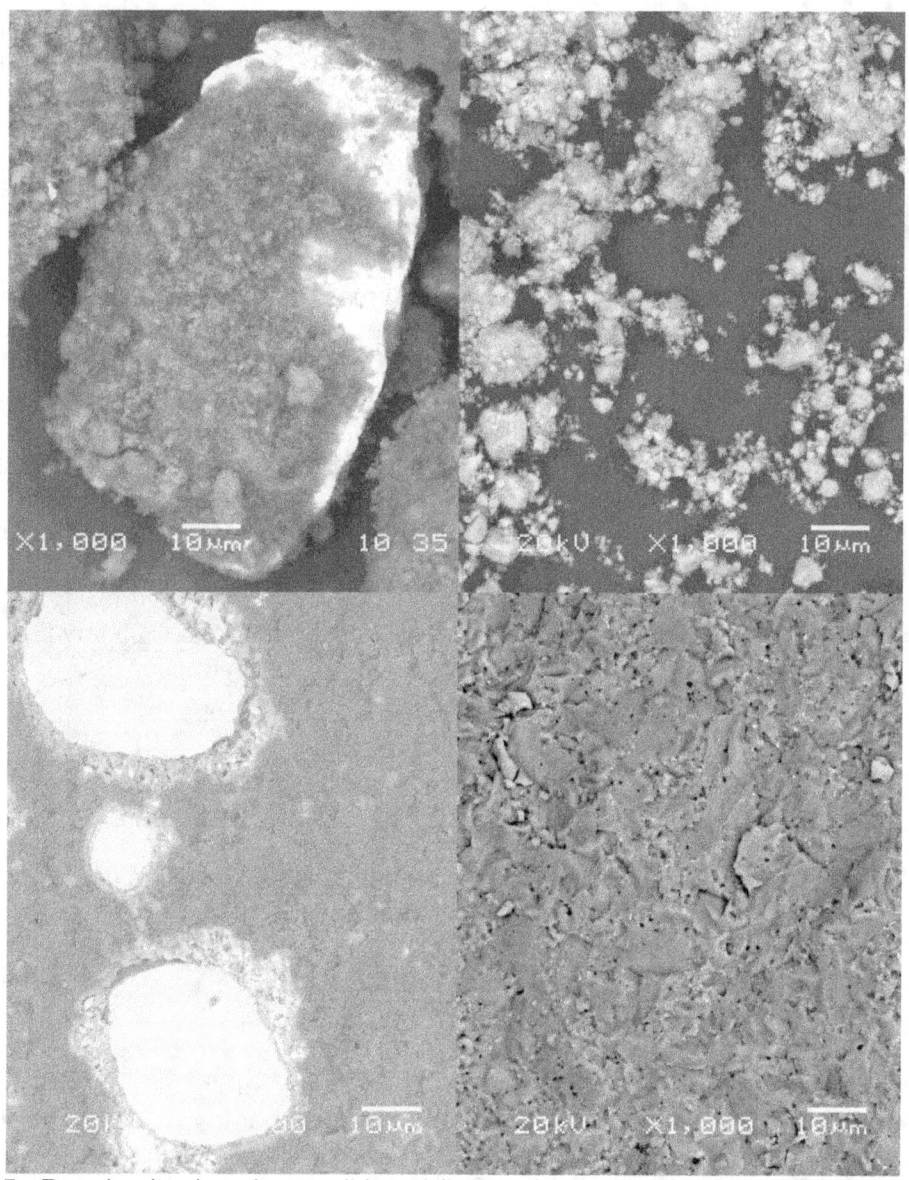

Figure 1.3.25. Powder (top) and consolidated (bottom) samples of TB00 (left) and TB100 (right) with Co-Mn. All images are backscattered (compositional), bright phase is Co-Mn where visible. Note the absence of large Co-Mn regions in the TB100 sample on the right.

Media-less milling studies of TiB_2 (TB100) and Co-Mn with media were found to result in effective comminution of the atomized binder powder (2hrs, ¼" media). The addition of $AlMgB_{14}$ was then accomplished by way of a subsequent step after the complete comminution of the binder by TiB_2. The resulting size and spatial distribution of binder was found to be uniform throughout the powder. This approach uniformly incorporates the binder in the mixed-phase boride composition. This was a definite improvement over earlier methods that resulted in an inhomogeneous distribution of micron-sized binder particles that did not achieve the desired improvement in toughness. The new approach may deliver results that approach the industry-standard microstructure seen in cemented carbides. This should provide a significant improvement in fracture toughness with only a minimal reduction in hardness.

The overriding objective is to achieve a uniform distribution of binder through the bulk of a mixed-phase composite, without appreciable agglomeration and reactivity with the matrix during consolidation at high temperature (e.g., 1300° to 1600°C). Adding the ductile binder to boride powder can in principle be accomplished through several approaches, not all of which produce favorable results. Simply mixing the gas-atomized powder with a charge of mixed-phase boride results in an inhomogeneous cermet, with no improvement in toughness. The challenge in mixing is due to the extreme ductility of the binder (> 40% elongation at failure).

Samples were prepared with this improved mixing approach for consolidation at Carpenter. As mentioned below, this approach appears to have produced a breakthrough in the quality of the $AlMgB_{14}$-TiB_2 cermets. The addition of Co-Mn is also being investigated with the anticipation that a binder phase would shorten the sintering time and enable lower temperature, thus reducing costs and energy intensity for production quantities. Lower sintering times and temperatures both help to (1) minimize reaction between the binder and the matrix, and (2) minimize grain growth that can be detrimental to wear resistance. Table 1.3.4 lists samples that were consolidated by Dynaforging. A corresponding set of samples were also hot pressed at Ames for comparison under conditions listed in Table 1.3.5.

Table 1.3.4: Samples consolidated at Carpenter Powder Products (CCP) by Dynaforging.

Composition	Co-Mn fraction
$AlMgB_{14}$ + 60 vol% TiB_2 (Z-TB60)	4.5 vol%
TiB_2 (TB100)	4.5 vol%
$(Ti,Zr)B_2$ (TZB)	4.5 vol%
WC	20 vol%

Table 1.3.5: Consolidation conditions for samples with addition of Co-Mn.

	Ames	CCP
Time (minutes)	30	1
Temp (°C)	1400	1200
Pressure (MPa)	105	825

Due to reactivity between the boride powder and Co-Mn during prolonged dwell times associated with conventional hot pressing at temperatures higher than 1200°C, pre-reacted $AlMgB_{14}$ was investigated instead of the elemental powder mixture. This sample was otherwise identical to the TB60 sample listed in Table 1.3.4. The sample with eTB00 exhibited improved mixing between the TiB_2 and $AlMgB_{14}$ phases. However, due to the incomplete mixing in the rTB00 sample, agglomerates of TB00 were not completely densified. The rTB00 phase itself has limited sinterability due to the absence of reaction-sintering that occurs in the eTB00 material, yet is expected to have reduced reactivity with the Co-Mn binder phase. Figure 1.3.26 (top) shows differences in mixing, the lower portion of the images shows the differences in the distribution of Co-Mn phase in each composite.

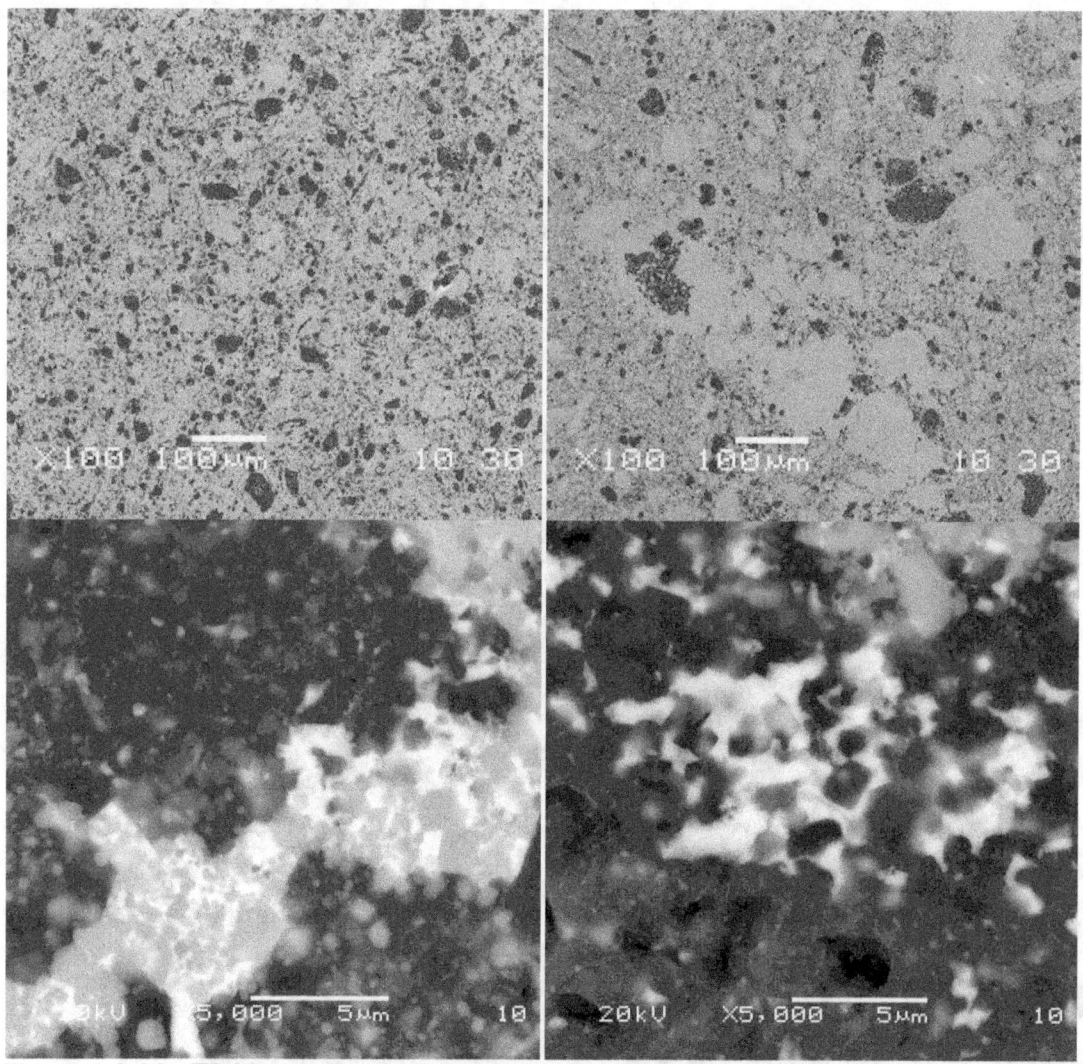

Figure 1.3.26: TB60 samples hot pressed at Ames containing Co-Mn. Images on the left correspond to as-milled AlMgB$_{14}$ while images on the right correspond to pre-reacted AlMgB$_{14}$ (dark phases). Bottom images highlight typical regions containing the Co-Mn phase (bright phase).

It can be seen that the Co-Mn addition indeed facilitated densification, as expected. The pre-reacted AlMgB$_{14}$ agglomerates were not enriched with Co-Mn during mixing, yet their outer regions are adjacent to Co-Mn-rich material. The binder phase can be seen in Figure 1.3.27 to have diffused into the agglomerates, reducing porosity. Samples processed with the improved mixing of Co-Mn do not typically exhibit large (e.g., tens of microns in size) agglomerates of the binder. At the standard hot pressing temperature (1400°C) some degree of liquid-phase sintering is involved, which increases the likelihood of a reaction. At the lower temperature involved in Dynaforging, reactivity can be minimized while the significantly higher pressure should enhance densification. It should be noted that while a shorter time for Dynaforging is noted in Table 1.3.5, this is time *at pressure*, not counting the time to reach temperature or the dwell time before application of pressure. Equalizing the temperature of the pressure-transmitting medium (i.e., the Pyrex glass) takes considerably longer and must be accomplished before pressure is applied. In addition to the possibility of unwanted reactions during this stage, initial sintering is occurring without applied pressure thus creating a network of bridged particles that reduce the effectiveness of the high pressure. The samples listed in Table 1.3.4 were processed at CPP and returned to Ames for evaluation, and a photograph of the samples after Dynaforging is shown in Figure 1.3.28. Initial results are promising. Each of the four samples was fractured with a hammer. (The pieces, including the mixed-phase boride, could not be broken by hand, in contrast to previous samples that were not fully dense.) Inspection of the fracture surface revealed a smooth texture with no visible porosity. Inspection with an optical microscope also showed no evidence of porosity throughout the volume of each piece. Unlike baseline boride composites without a binder, these pieces were highly resistant to fracture. Since the consolidation conditions were identical to previous samples, it appears that the technique of adding Co-Mn to TB100 powder and then adding the TB00 powder is the preferred route. Previous batches of boride powder consolidated at Carpenter were not at this level of homogeneous mixing of the binder phase.

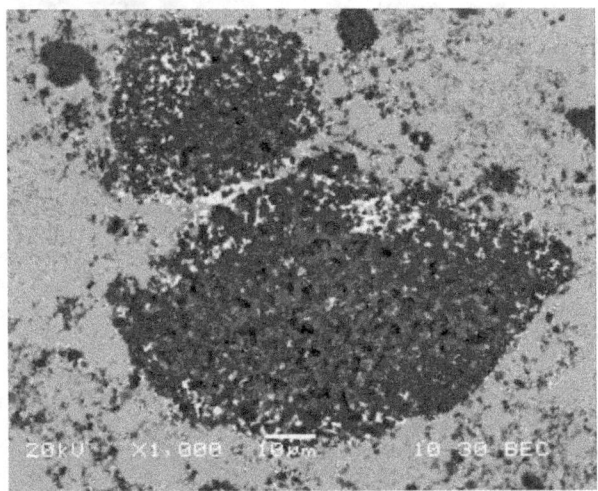

Figure 1.3.27: pre-reacted $AlMgB_{14}$ agglomerates (dark) with Co-Mn (bright) binder-assisted densification around their perimeters.

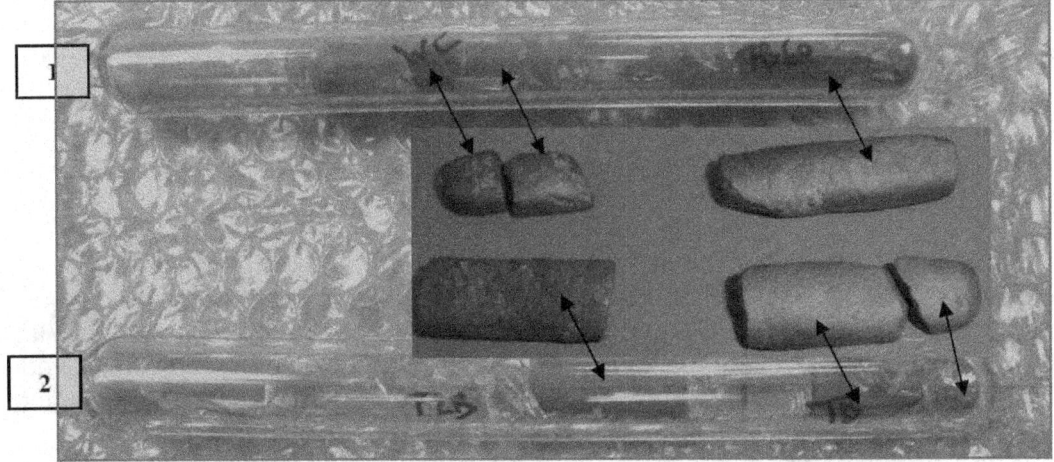

Figure 1.3.28. Samples of boride composites with Co-Mn after Dynaforging. The fractures seen in two of the samples were present before densification. These are the highest quality Dynaforged samples produced by the end of the project.

During the initial portion of this project, additions of Co-Mn binder had processing deficiencies associated with incomplete comminution of the ductile phase; most of these deficiencies have since been resolved. Previous improvements in distribution were due to comminution of the ductile Co-Mn with abrasive TiB_2 particles plus media. Small (1/4") media were used because the larger media used in earlier experiments caused the Co-Mn to adhere to the media and vial and fail to mix with the powder. Within regions where the binder phase was present, it clearly aided sintering and dramatically improved toughness, as expected. This was established through analysis of samples pressed at 1400°C at Ames Lab. Since this temperature results in liquid phase sintering and partial reaction with the boride phases, identical samples were sent to Carpenter (CPP) for lower temperature consolidation by Dynaforging. Higher pressures are achievable through the Dynaforge process compared with conventional hot pressing, which facilitates retention of the nanosize grains. The advantages of Dynaforging include reduced reactivity and grain growth due to the lower temperatures. Simple hammer tests demonstrated that the improved mixed-phase boride samples with Co-Mn binder addition were substantially more resistant to fracture than any previous samples, regardless of the preparation route. Figure 1-1 shows a typical fracture microstructure of one of the TB60 samples with ~5 vol.% Co-Mn consolidated by Dynaforging.

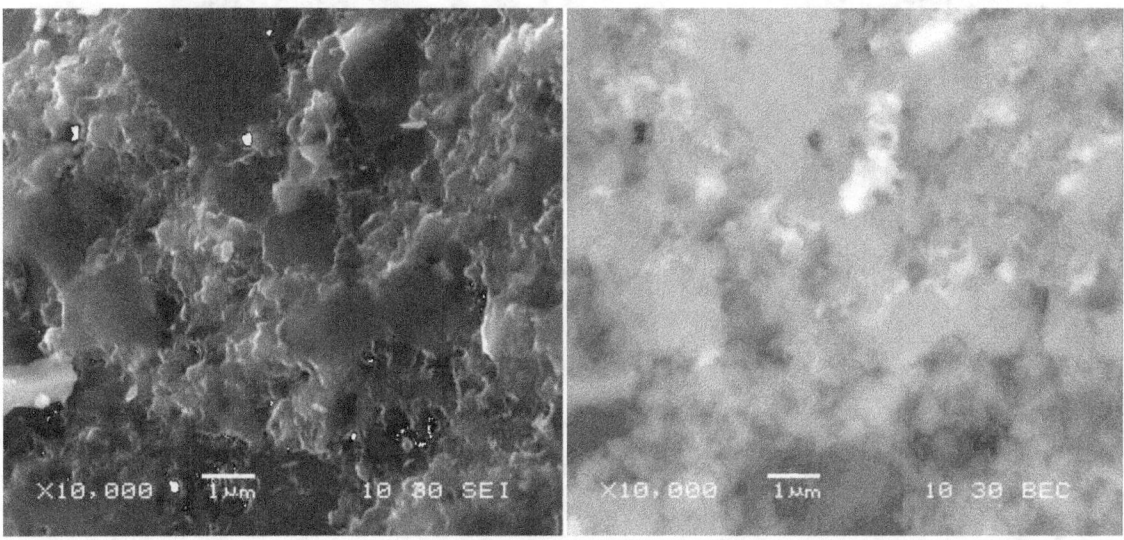

Figure 1.3.29: Nominally 40 v% AlMgB$_{14}$ (dark) - 60 v% TiB$_2$ (grey) with the addition of ~5 v% Co-Mn (bright) consolidated by Dynaforging. Fracture surface viewed topographically (left) and compositionally (right).

The upper portion of Figure 1.3.29 shows a region of excellent mixing of the bright Co-Mn phase acting as binder between the larger TiB$_2$ grains. The results are quite promising in that Co-Mn additions appear to facilitate densification at temperatures as low as 1200°C. Shown more clearly in Figure 1.3.30, the Co-Mn distribution in most regions of this sample is quite uniform, with little or no agglomerates. The only observable porosity exists within the grains.

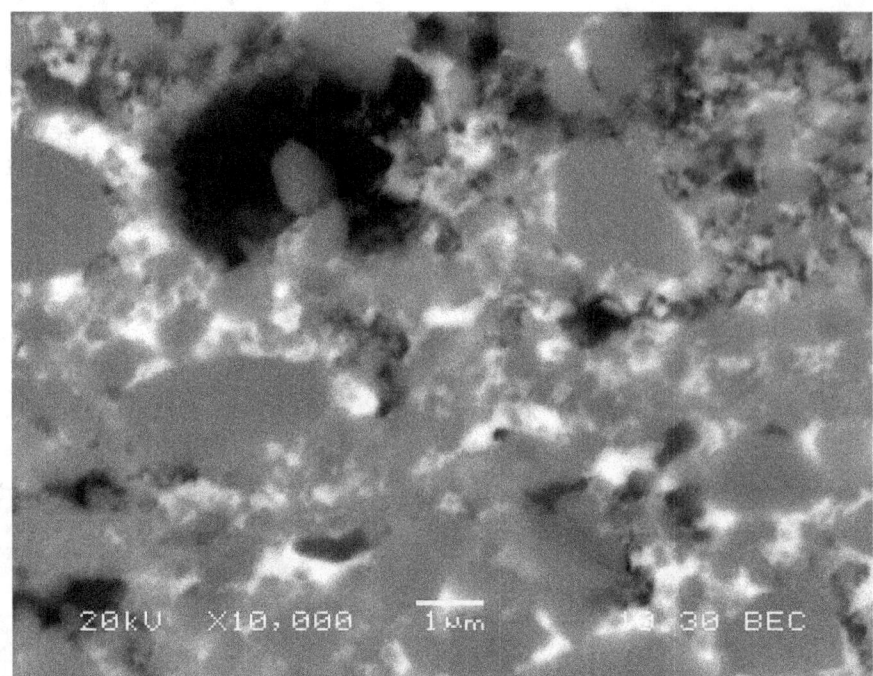

Figure 1.3.30: Polished cross section of sample in Figure 1.3.29. Note the excellent distribution of the binder phase (bright contrast) around the boride grains. Also, the only observable porosity occurs within the grains themselves.

1.4: Oxygen Reduction

When oxygen is present during powder processing, $MgAl_2O_4$ (spinel) forms in the boride nanocomposite. This is particularly problematic for large-scale processing because control of oxygen becomes increasingly difficult and costly as the size of the handling and processing environments increases. Analysis of consolidated compacts prepared at Ames Laboratory and by NewTech Ceramics reveal spinel contents ranging from a few percent to over 10 percent by volume. Such levels of spinel significantly reduce the hardness and wear resistance of the material; consequently, we investigated cost-effective solutions to reduce or eliminate the spinel phase in boride nanocomposite compacts.

Figure 1.4.1 illustrates how TiB_2 grains with surface oxides can degrade composite properties. Layers of $AlMgB_{14}$ and TiB_2 powder were hot pressed to form a couple in which cracks were induced across the boundary by a Vickers indenter. In most cases, the cracks travel across the boundary and show good bonding between the two phases. In Figure 1.4.1, a crack initiated in the $AlMgB_{14}$ (dark) phase is deflected at the boundary with the TiB_2 (light) phase. At this location there is a large single-crystal grain of TiB_2 of the size and morphology of as-received Aesar TiB_2. Past experiments showed that as-received TiB_2 particles sinter poorly due to their large size and the ubiquitous surface oxide layer. Milling in inert atmosphere exposes atomically-clean surfaces and makes the powders much more readily sinterable [Peters09]. In the figure, the crack is seen to follow the oxide layer at the interface between the $AlMgB_{14}$ and the TiB_2. Even thin layers of oxide cause this delamination between grains.

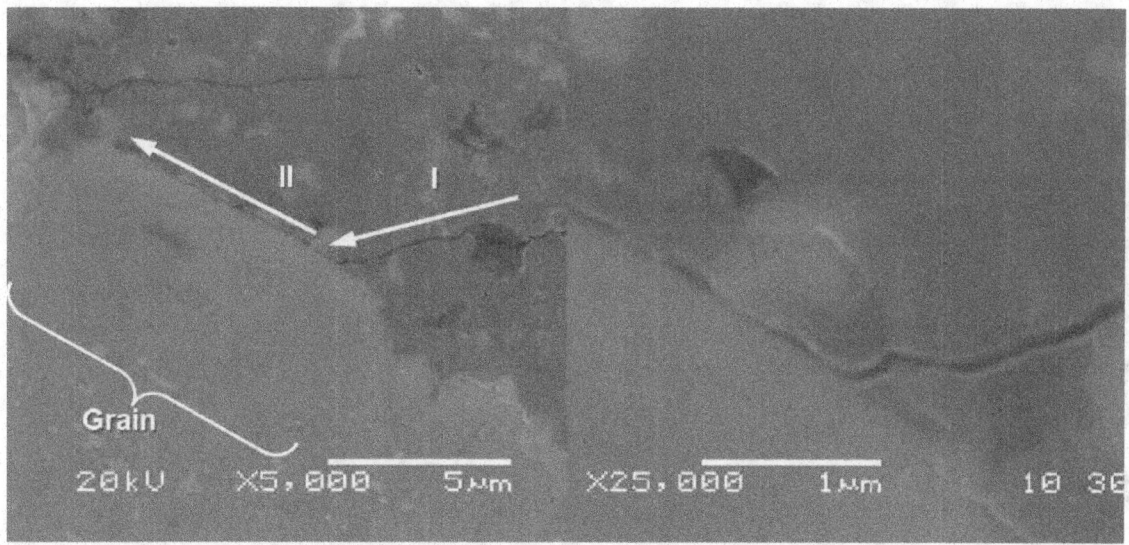

Figure 1.4.1. Layered TB00 and TB100 couple with induced crack showing failure along $AlMgB_{14}$ - TiB_2 boundary (labeled II). This type of failure is highly unusual in $AlMgB_{14}$-TiB_2 composites, and is due to the presence of surface oxides.

Oxygen removal from the precursor powders and/or the final composites is thus important for commercialization. The spinel phase that forms is typically as high as 10 vol%, a level that degrades the composites' wear resistance. This holds true for both bulk and coating forms of the material. The hardness of spinel is considerably lower than either the $AlMgB_{14}$ or TiB_2 phases, and spinel is also more brittle. For example, in a baseline $AlMgB_{14}$ sample (TB00) to which 30 vol% spinel was intentionally added during processing (approx. 37 vol% total spinel), hardness was reduced by 15% compared with the baseline material and the erosion rate nearly doubled, from 2.0 mm^3/kg for a typical (10 vol%) TB00 sample to 4.1 mm^3/kg with the excess spinel addition. The presence of oxides, particularly surface oxides on particles, can diminish the synergistic bonding between $AlMgB_{14}$ and TiB_2, which is a primary reason for the positive deviation from the rule-of-mixtures for hardness and wear resistance in these composites. When the interfaces between the two boride phases are free from oxides, crack propagation along the phase boundaries is not energetically favorable [Cook10]. Crack propagation readily occurs along the interface between the boride phases when a surface oxide is present.

As an example of the desirability of minimizing spinel within these composites, the sample containing 37 vol% spinel was prepared (again, versus ~10 vol% typically) for fracture analysis. Fracture of the specimen shows the primary failure mode as intergranular around the spinel grains, as shown in Figure 1.4.2 (bright). These mechanisms are also likely the cause of the elevated dry erosion rate for this sample, which was twice as high as similar baseline (TB00) samples.

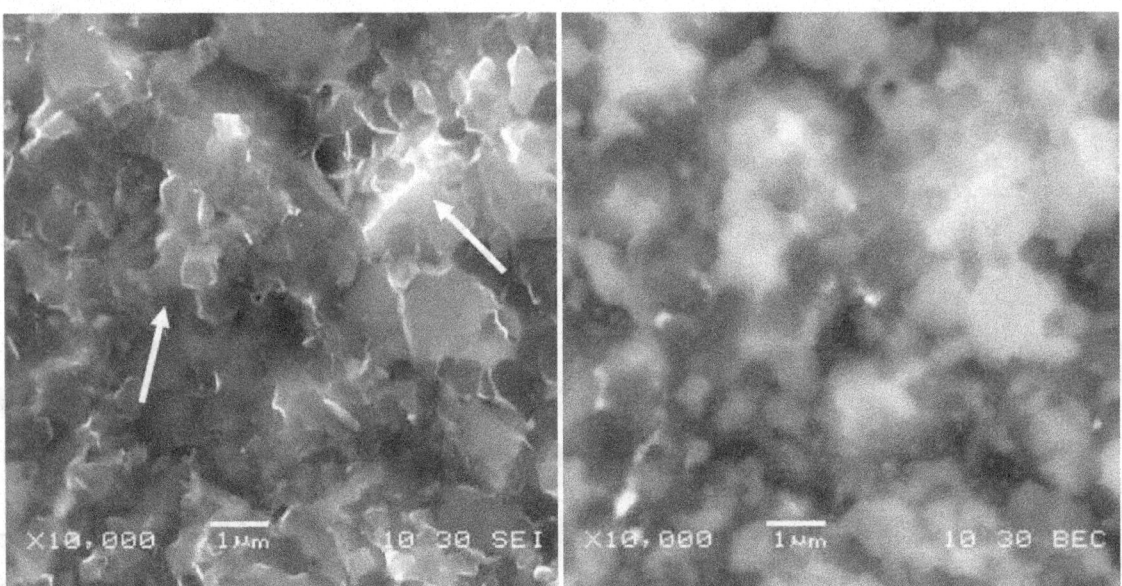

Figure 1.4.2: Fracture micrographs of $AlMgB_{14}$ intentionally prepared with 30 vol.% $MgAl_2O_4$ spinel, showing examples of intergranular fracture (arrow). Topological (left) and compositional (right) imaging with $AlMgB_{14}$ (dark) and spinel (bright) present.

Transgranular fracture is the primary failure mode in the single-phase and mixed-phase boride composites, again due to the strong bonding between the $AlMgB_{14}$ and TiB_2 phases [Cook10]. The presence of spinel between boride grains weakens the material by increasing the instance of less-desirable intergranular fracture, which reduces wear resistance. Multiple methods of O reduction have been employed, as described below. These efforts can be divided into three broad categories; powder washing, O gettering, and oxy-nitride formation. These may or may not be used in conjunction with the previously developed technique of vacuum outgassing . Further evaluation of oxygen contamination is discussed in Task 4 in reference to commercially produced powders.

A significant fraction of O in the composites is assumed to be present in the raw, as-received elemental precursor powders. This is especially true of the fine B powders typically used. One method of treating the as-received powder is vacuum outgassing. In general, this involves heating the powder high enough to release adsorbed and weakly bonded air and water molecules from the powder, which can be a considerable amount due to the high surface area of fine powders. These outgassed products are then pulled away by the vacuum pump. In the treatment of B powders, there is an additional mechanism available, due to the low melting and high volatility of B_2O_3. Above the melting point of the boron oxide (>500ºC), its vapor pressure is high enough for it to be evaporated under vacuum, thus cleaning the powder surfaces. This has been the typical method of preparing boron powder in Ames Laboratory, but it has not been effective at removing all of the oxygen. It is possible that during heating, enough of the oxygen present reacts with impurities in the B (i.e. Mg) to form stable oxides that cannot be removed by heating in vacuum.

Powder Washing
Another property of B_2O_3 that makes boron amenable to other possible cleaning techniques is that the oxide reacts with water to form boric acid ($B(OH)_3$) which is then soluble in water. Additionally, B_2O_3 can react with methanol to form a volatile compound that can be removed by vacuum without the heating required to vaporize the oxide alone.

Most commercially available B powders contain a significant quantity of oxygen, generally in the form of B_2O_3. When present in the $AlMgB_{14}$ composition, oxygen will tend to form $MgAl_2O_4$ as a stable second phase. One approach tested for removal of the B_2O_3 from B powder was to wash the material with methanol. During this process, the methanol complexes with the B_2O_3 to form a soluble organic compound that evaporates along with the methanol when the treated powder is dried.

The experiments to evaluate pre-treatment of B powder with methanol to remove oxygen impurities from the powder have indeed shown beneficial results. Amorphous B powder (95% purity metals basis) exhibited a weight loss of 12.5% after methanol washing. This corresponds to an oxygen loss of 8.62 wt.% or 9.70 wt.% depending on whether the oxygen is combined as B_2O_3 or H_3BO_3, respectively. High-purity crystalline B (99.5% metals basis) showed a weight loss of 2.5% after washing. Fired batches of blended powders having a nominal composition of $AlMgB_{14}$ showed a reduced tendency to form $MgAl_2O_4$ when the washed B powders were used. For typical batches, the $MgAl_2O_4$ content was reduced by ~15% while the $AlMgB_{14}$ content was increased by 20% or more.

In addition, boron powder washed with methanol at ORNL has been used in the preparation of $AlMgB_{14}$ at Ames. Vacuum outgassing of as-received powder at 1100°C is effective at removing some oxide and adsorbed gasses, yet there is room for improvement. SEM examination of $AlMgB_{14}$ powders milled using each boron source showed no morphological differences. At the same electron beam settings, the powder from the methanol-treated source indicated less charging from the beam, which indicates higher conductivity due to lower oxide contamination as seen in Figure 1.4.3. This is a very qualitative measurement and needs to be reinforced by actual chemical analysis. EDS results were inconclusive, and the technique is not optimal for determining B or O, especially in B-rich solids.

Figure 1.4.3: $AlMgB_{14}$ powders from different boron sources show different degrees of charging (brightness) under identical settings.

As presented in Figure 1.4.4, erosion results for baseline $AlMgB_{14}$ produced with Oak Ridge National Labs (OR) boron compared to a standard Ames Lab baseline sample were 2.05 mm^3/kg and 2.79 mm^3/kg, respectively, which indicates that spinel concentrations were effectively reduced in the OR sample. Note that these samples did not contain the TiB_2 reinforcement phase which significantly improves wear properties. Previous Ames Lab samples with 60 vol% TiB_2 have shown rates below 1 mm^3/kg.

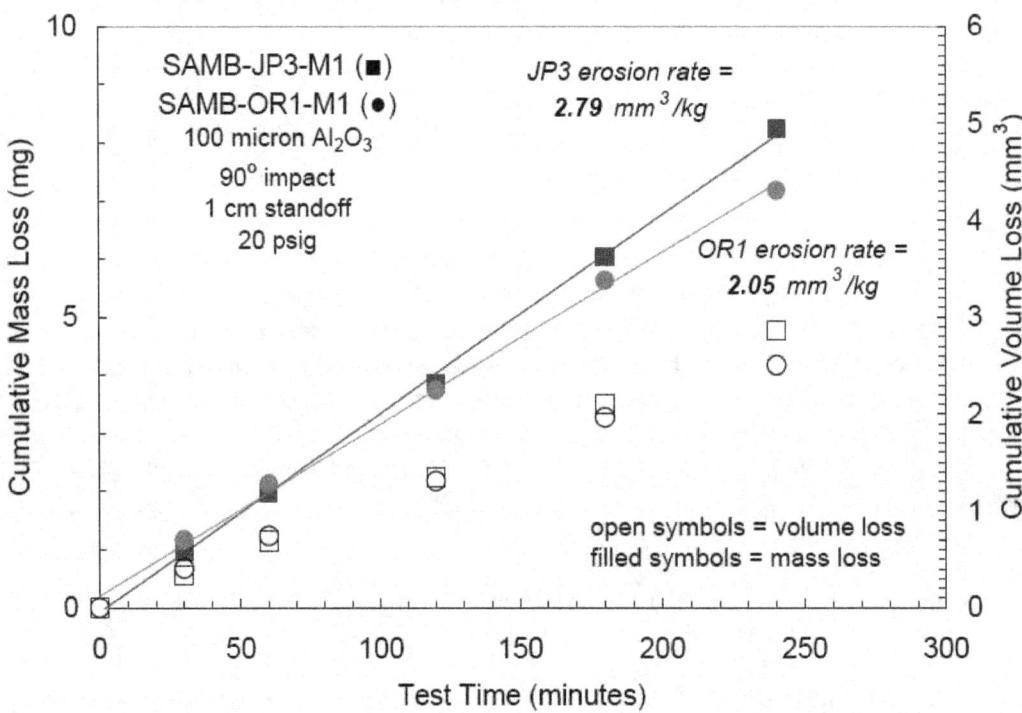

Figure 1.4.4: Erosion losses for baseline samples made with Oak Ridge treated boron (OR) and standard Ames Lab baseline samples. Note listed averages, OR sample experienced a higher flow rate of erodent giving the illusion of similar erosion rates.

Samples of 60 vol% MM TiB$_2$ composites were produced with methanol-treated boron from Oak Ridge (OR), but are not pictured due to the fact that the O-rich spinel phase cannot be resolved by SEM in the composite material.

Oxygen Gettering

Treatment of precursor powders does not eliminate contamination that may be introduced during milling and subsequent processing. This is important in industry where preservation of oxygen-free atmospheres during all stages of processing becomes increasingly difficult and expensive. One approach may be the use of oxygen getters during powder processing. Our initial research on the baseline AlMgB$_{14}$ and related systems included small additions of silicon, which was found to increase hardness [Cook00]. At that time the mechanism responsible for increased hardness was not fully understood, although substitution of Si on the Al site had been examined elsewhere [Kolpin08,Lowther02]. Band structure calculations based on a full-potential, linear augmented 3-plane wave method revealed a possible shift in the position of the Fermi level within the band gap of AlMgB$_{14}$, which could be favorable for increased hardness [Harmon02]. Recently, we have speculated that the presence of silicon could also have acted as an oxygen getter, reducing the amount of spinel in favor of a more benign secondary phase such as SiO$_2$. Two test samples were synthesized: one in which Si was substituted for 50% of the Al (i.e., Si substitution), the other in which the same quantity of Si was added in addition to the standard stoichiometric ratios of AlMgB$_{14}$ (Si addition). In either case, the effect should be reflected in an x-ray diffraction pattern of the hot pressed compacts, which should reveal if substitution or gettering is the dominant mechanism. Reduction in intensity of spinel peaks would confirm oxygen gettering and would also be accompanied by reduced AlMgB$_{14}$ peaks in the substitution sample as there would be insufficient Al for complete 1:1:14 formation. A shift in peak position in the 1:1:14 peaks would indicate Si substitution. The samples pressed at 1400°C showed a slight decrease in intensity of the spinel peaks compared with a control sample without Si addition, although the total amount of Si was relatively low. In order to confirm the oxygen gettering effect, a second set of samples containing 5 times the amount of Si as previously was prepared.

The addition of Si or Mn is based on past research that showed a slight increase in hardness with the addition of Si. As Mn is not expected to substitute into the 1:1:14 phase, only addition samples were prepared, based on an equal atomic fraction of Mn and Si. Instead Mn additions had a secondary possibility of forming MnB_2, which may behave beneficially as an alternative to TiB_2

X-ray diffraction analysis of Si containing substitution samples revealed a reduction in intensity of the 1:1:14 peaks, indicating that substitution did not occur, and less Al was available for the formation of $AlMgB_{14}$. The addition samples exhibited no apparent reduction in spinel peaks, yet in all Si containing samples additional phases appeared. These additional peaks were indexed as FeSi and $FeSi_2$, with the addition samples containing more FeSi, while the substitution samples were richer in $FeSi_2$. FeB peaks were also reduced or eliminated. Total Fe contamination of the Si samples also appeared to be higher, as indicated by an Fe-rich melt or "squeeze-out" from the samples during hot pressing. Fe and/or FeB have been determined to facilitate sintering, yet despite the higher Fe content, Si containing samples did not appear to densify as well as standard baseline samples. This would indicate that the iron silicides are not effective sintering aids. Sintering at 1500°C improved densification, with one Si addition sample showing an improved erosion rate of 1 mm^3/kg, compared with a value of 2 mm^3/kg for standard baseline samples. The excessive Fe contamination is likely due to the abrasive nature of the Si additions.

In addition to its possible role in gettering oxygen, Mn may also react with boron to form MnB_2. MnB_2 has been reported to possess a bulk modulus comparable to TiB_2, based on structure calculations [Vajeeston01], and may offer a suitable addition or alternative to TiB_2 for certain applications. Additives of rare earth elements are also of interest, for example scandium, in the form of ScB_2 or $ScMgB_{14}$ has recently been reported to possess desirable properties of high hardness and modulus [Kolpin08,Levchenko06]. ScB_2 possesses a lower melting temperature than TiB_2 (2250 vs 3225 °C) and therefore may be useful for reducing sintering temperatures. $ScMgB_{14}$ has also been reported, based on theoretical calculations, to have a larger (negative) energy of formation, and therefore stability, compared with $AlMgB_{14}$.

Unfortunately, neither the Si nor Mn additions were found to yield a substantial reduction in spinel. Si additions formed FeSi or $FeSi_2$, and Mn additions formed mixed (Fe,Mn)B. If successful, solid-phase gettering would still have resulted in a new oxide phase incorporated in the material, albeit with less degradation of the $AlMgB_{14}$ phase. An alternate approach was discussed in which residual O is converted during high-temperature processing to a gaseous oxide (i.e. CO/CO_2 or H_2O) which could then be removed during densification, leaving no residual phase. While spinel is a highly stable phase, preliminary calculations have shown that under sufficient vacuum, C should be able to reduce the oxide in the presence of $AlMgB_{14}$. Reduction of oxides by C in TiB_2 ceramics has been reported [Baik87,Finch86].

Consequently carbothermal reduction of spinel by addition of elemental carbon was investigated. Preliminary thermodynamic calculations indicated that the reaction of carbon and oxygen at 1500°C would be favored. Based on an estimate of 10 vol% spinel in a 1.3 g $AlMgB_{14}$ sample, 0.05 g C was determined to be sufficient to react with the O present (by forming CO, which is stable at the hot pressing temperatures). Comparison of the same series of $AlMgB_{14}$ with and without C addition showed essentially equal fractions of spinel by XRD, although comparison of peak height is at best an imperfect technique. Also, the $AlMgB_{14}$ peaks appeared roughly half as intense after C addition, indicating that excess C may be detrimental to the formation of the $AlMgB_{14}$ structure. SEM revealed that both samples were microstructurally similar, suggesting that no significant reduction in spinel took place. As discovered previously, C additions appear necessary for full densification of TiB_2 (TB100) at low temperatures (1400°C) [Peters 2009]. It was also found that for samples a few mm thick, C diffusion from the graphite die was sufficient to aid densification. Thus it is likely that there is already C diffusion into the $AlMgB_{14}$-containing samples during conventional hot pressing. It is possible that this carbon may be inhibiting $AlMgB_{14}$ formation, as with the intentional C additions.

Previous studies on the sintering behavior of $AlMgB_{14}$ indicated that lines attributable to spinel do not appear in the x-ray diffraction patterns of samples sintered below 900°C. The carbothermal process has been calculated for the reduction of spinel, but since O is present in the initial powder in multiple forms of oxide layers this process may be more amenable to O reduction in pre-reacted powders in which the O has already converted to spinel. Hydrogen reduction furnaces, which are used in many industrial processes for O removal, form H_2O vapor as the byproduct. While not thermodynamically favorable for the reduction of spinel, H reduction may be possible below its formation temperature (900°C).

Oxy-nitride Formation
A different approach was taken in another series of experiments with the objective of converting the existing spinel phase to a more benign phase, rather than seeking to remove it altogether. Spinel is moderately hard (15 GPa) yet highly brittle, like many other oxide ceramics, with a K_{IC} value of 1.7 – 2 MPa√m at room temperature. On the other hand, SiAlON is an oxynitride ceramic that has a higher toughness (5 -7 MPa√m) and is known to accommodate other cations (such as Mg). In advanced cutting tools, for example, SiAlON is regarded as less detrimental than spinel. In fact, at high temperature encountered during machining, the high-temperature hardness of SiAlON exceeds that of alumina and WC, as seen in Figure 1.4.5.

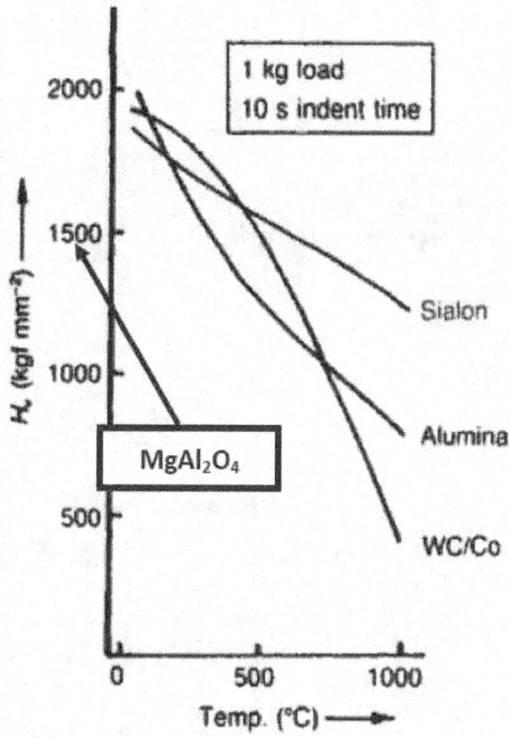

Figure 1.4.5. High temperature hardness of SiAlON, Al_2O_3, and WC [Smallman99]. Also shown is the room temperature hardness of $MgAl_2O_4$ spinel.

Reactions between Si_3N_4 and Al_2O_3 or $MgAl_2O_4$ have been documented [MacKenzie96] and thus the addition of Si_3N_4 to $AlMgB_{14}$ powders may provide a cost-effective route to formation of a tougher SiAlON phase. In preliminary experiments silicon nitride and spinel were milled and subsequently hot pressed at 1400° and 1500°C for one hour. No reaction between the two was found by XRD, suggesting the reaction will likely require a higher temperature to overcome sluggish kinetics. Samples were then heat treated up to 1800°C to find a suitable reaction temperature. This is well within the temperature range of some consolidation efforts, including Greenleaf's. Additionally the ability to handle powders in air is highly desirable for most industrial applications. With the addition of silicon nitride, the issue of spinel formation as a consequence of air handling may be less of an issue for the composite materials.

Reaction sintering of spinel with Si_3N_4 at higher temperatures (up to 1800°C) was performed by Greenleaf. This approach may be of particular benefit for two reasons; commercial powders from NewTech have been shown to be high in O, and machining applications for the boride ceramics require the highest possible toughness to prevent catastrophic failure of the bulk cutting tool. In the future, it may be possible to toughen the bulk material with the addition of both SiC whiskers and Si_3N_4. Both materials could be added to the commercially available powders prior to hot-pressing.

In similar hot pressing studies (Task 2), 1600°C was identified as an acceptable consolidation temperature at Greenleaf using their applied pressure of ~30 MPa. Figure 1.4.6 shows the reduction in intensity of spinel peaks at 1600°C and above with the corresponding formation of β' SiAlON. The phase that appears to form is more complex than the reference β' SiAlON due to the inclusion of Mg [MacKenzie96]. Spinel appears almost completely eliminated at 1800°C. However, there may have been insufficient Si_3N_4 to react with all of the spinel at the highest temperature. If spinel removal is proven to be reproducible with $AlMgB_{14}$ + TiB_2 additions, this

71

may constitute a significant improvement for commercially-prepared composites for which ultra-high-purity processing environments are not cost effective.

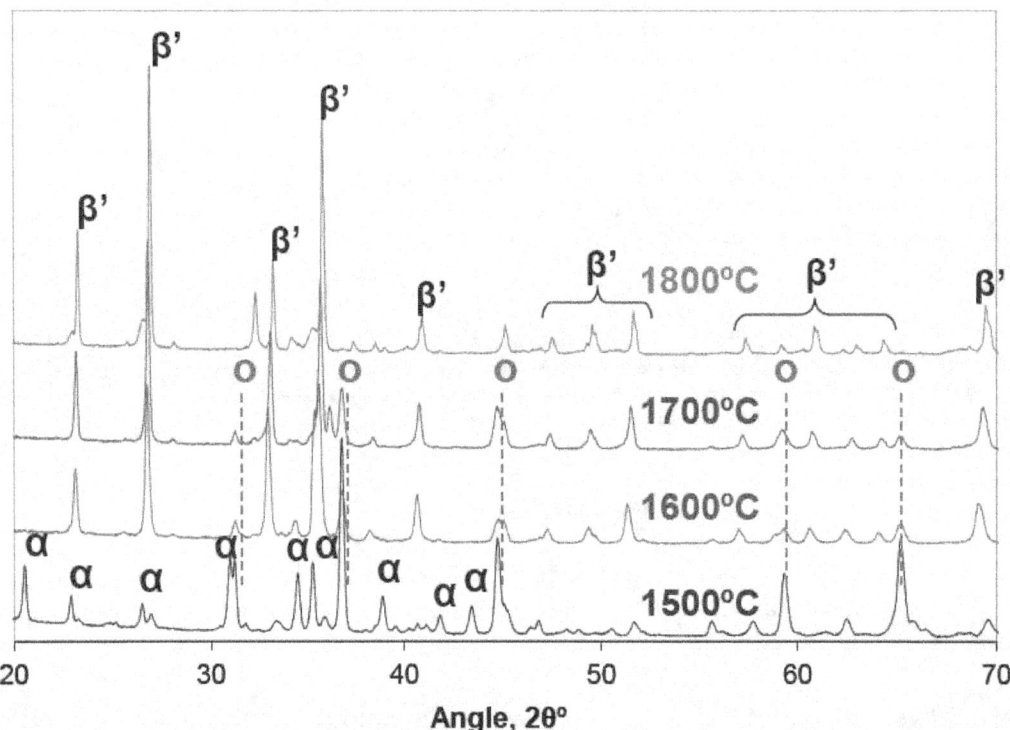

Figure 1.4.6: XRD patterns of Si_3N_4 + $MgAl_2O_4$ hot pressed at 1500 to 1800 °C. Major peaks of α-Si_3N_4, $MgAl_2O_4$ (**O**), and β'-SiAlON are labeled.

It has been shown that intergranular fracture is the primary failure mode in the single and mixed-phase boride composites. The presence of spinel between boride grains weakens the material and reduces wear resistance. Future efforts to fully develop the processing technology of in-situ conversion of Si_3N_4 and $MgAl_2O_4$ into SiAlON appear highly desirable.

Task 2: Development of Scale-up Technology

Next-generation powder processing methods were developed, with the objective of increasing abrasive and erosive wear resistance in bulk composites by improving fine-scale mixing of the matrix and reinforcement phases prior to consolidation. Earlier powder processing approaches were based on high-energy ball milling (HEBM), followed by hot uniaxial or isostatic pressing. Improvements to this method involved the use of chemical additives such as stearic acid and methanol, to promote particle size reduction and more effective blending of the phases. Initial samples prepared with stearic acid have shown improved erosion resistance, along with a substantial reduction in milling duration compared to conventionally-prepared samples. Efforts to reduce milling time also consisted of blending the elemental powders, followed by cold isostatic pressing (CIP) and reaction sintering to form the $AlMgB_{14}$ phase. Sintering temperatures as low as 1100°C were found to result in formation of the desired phase. Utilization of the CIP/sinter technique to form the alloyed powder will enable scale-up of powder batches with minimal milling. In both cases the resulting powder was finer and possessed better flowability than samples prepared in the conventional dry milling procedures.

Industrial hot pressing is generally scalable to keep up with increased powder production. Still, this technique was compared with alternative forms of consolidation such as spark plasma sintering (SPS) and a proprietary rapid consolidation process known as Dynaforging. As discussed below, the advantages and disadvantages of each lend themselves to different formulations of the $AlMgB_{14}$-TiB_2 composite family.

2.1: Powder Processing

Various raw material powders were obtained for preparation of $AlMgB_{14}$ samples and $AlMgB_{14}$ – TiB_2 composites. Both amorphous and crystalline boron (B) was evaluated. The amorphous B is lower cost, but has relatively low purity (~95% metals basis) and contains oxygen. Large scale powder production is necessary for industrial applications. At the beginning of this project, small scale (~2 gm) samples were prepared by high energy SPEX vibratory milling. As described below, numerous alternate synthesis techniques were investigated for the possible application to industrial scale powder synthesis.

Microwave Heating

Microwave heating has been used to produce nanopowders in a variety of materials systems. Most of the work has been done in oxides and metals The advantages of microwave heating are lower processing temperatures and faster heating rates. These attributes should help to maintain nanosized particles. Heating of loose powders of the mixed constituents is done in a microwave transparent crucible (i.e. BN) at temperatures sufficient for the reactions to occur. The reaction product is a friable powder compact that can be easily milled to obtain fine and nanoscale powders. The microwave heating process has the capability for scale-up to commercial quantities and requires lower energy milling to process the powders.

Microwave heating was evaluated for converting the blended constituent powders to $AlMgB_{14}$. Samples were heated in a microwave furnace under an Ar atmosphere. The initial powder batch, which included methanol-washed B, utilized dry ball milling under Ar in a polyethylene container with WC media. The milled powder was contained in a microwave transparent reaction bonded silicon nitride crucible. Experiments showed that the powders coupled readily to the microwaves and quickly heated to the reaction temperature. Only about 500 watts of power was needed to heat 5g of powder to ~1200°C. The resulting product, which was in the form of powder, was primarily AlMgB14 (>70%). Some expected secondary phases, including $MgAl_2O_4$, were detected by X-ray diffraction analysis. The particle size of the reacted powder was measured and was found to be at the high end of the size distribution of the original constituent powders. This suggests that it will be possible to retain an ultra-fine particle size, if nano-sized starting materials are used in the initial powder blend. Some powder samples were also heated in a graphite sintering furnace, again using an Ar atmosphere and a reaction bonded silicon nitride crucible to hold the powder. The results from these tests were similar to those observed with the microwave-heated samples. These results are encouraging for continuation of scale-up and consolidation to form dense parts.

Slurry Milling
Powders were milled in methanol similar to the powder cleaning technique of methanol washing. This was attempted to potentially combine the O reducing washing step with milling, with the added benefit of the liquid lubricant minimizing agglomeration and thus improving the homogeneity of the powders. This so called "slurry milling" is also amenable to producing homogenous powders in larger batches by keeping the constituent powders evenly mixed and distributed.

Two slurry processing methods were demonstrated for blending $AlMgB_{14}$ and TiB_2 powders together to form a composite powder mixture at Oak Ridge National Laboratory. The first approach was attritor milling. Equal weights of pre-reacted $AlMgB_{14}$ and TiB_2 powders were added to a Teflon-lined milling jar containing 3-mm diameter yttria stabilized zirconia media. Initial slurries utilized methanol as the solvent. The slurries were dried following milling and 55-65g of powder were reaction fired by either radiant heating in a graphite resistance furnace or microwave heating. The radient heated reacted powder was analyzed by X-ray diffraction and was found to consist primarily of $AlMgB_{14}$ with small amounts of $MgAl_2O_4$ (spinel) and AlB_{31}. For microwave heating, the blended powders were held in an aluminosilicate crucible which was surrounded by an insulation pack. Coupling of the powders with the microwave energy is excellent and, in this test, the temperature quickly shot to >1400°C before the power was reduced. It is likely that the rapid temperature excursion resulted in vaporization loss of Mg from the mixture. Subsequent analysis of the reacted powder showed that it had been converted to AlB_{12}; no other compounds were seen in XRD. Slurry milling was also performed with isopropyl alcohol to form a slurry that contained approximately 12-15 vol.% powder. A high density polyethylene stirrer was used to agitate and mix the slurry. The attritor was operated at medium speed for 1 hour to complete the blending. The slurry was then poured through a sieve into a glass tray and placed in an oven at low temperature to dry. After drying, the powder cake was broken up using a mortar and pestle and then poured into a plastic jar for storage.

The second slurry processing method that was investigated was wet jar ball milling. A high density polyethylene jar containing WC milling media was used for this method. Again, equal weights of $AlMgB_{14}$ and TiB_2 were poured into the milling jar along with sufficient isopropyl alcohol to obtain a 12-15 vol.% slurry. The jar was rolled at medium speed on a ball mill rack for 4 hrs. to blend the powders. As with the attritor milling method, the blended slurry was poured through a sieve into a glass tray and placed in an oven at low temperature to dry. In both processing methods, 120-150g of powder was used to demonstrate the process. The dried powders were easily die-pressed to form a pellet having a green density of greater than 50%. Either of the two approaches can be used to obtain uniform blends of the constituent powders and both methods can readily be scaled to kilogram quantities or larger for commercial production.

Slurry milling of small powder batches was also conducted at Ames Laboratory in a laboratory scale vibratory mill. Planetary milled powder was utilized as a master quantity of $AlMgB_{14}$ with 60 volume percent TiB_2 powder suitable for slurry milling experiments. This powder production route was chosen due to the low Fe contamination typical of planetary milled powders as well as the capability of planetary milling to produce large powder quantities. One sample was prepared from this powder using conventional techniques, namely hot pressing of the dry milled powders. The slurry milled samples used the planetary milled powder as precursor material into which methanol was added. The slurry was then added to a steel vibratory milling vial with steel milling media. Milling was carried out for varying time periods followed by vacuum drying to remove the methanol, after which the powder was consolidated by hot pressing. Extremely fine particles were found to coat the milling media and vial walls. The last material to vacuum dry formed a "cake" within the vial. Particles within this "cake" had agglomerated upon drying to produce a firm compact. A second series of samples utilized smaller milling media, which generally facilitates particle size reduction. The qualitative observation that the particles resulting from slurry milling possessed the consistency of smoke is very encouraging and points to this technique as an effective means to achieve even more highly refined microstructures in the consolidated compacts. Slurry milling is intended to improve both the intermixing and size reduction of the reinforcement phase TiB_2 with the $AlMgB_{14}$ matrix.. These trials showed that the Mg in the planetary milled powder reacted with the methanol resulting in no $AlMgB_{14}$ formation following hot pressing. Processing parameters were adjusted to avoid Mg interaction with the methanol. Boron powder and commercial TiB_2 powder were slurry milled without Al and Mg powders. The Al and Mg were added after the pre-milling and vacuum drying of the B/TiB_2 mixture and all components were dry milled followed by hot pressing. While successful in forming an $AlMgB_{14}/TiB_2$ composite, the initial hot pressed compacts exhibited excessive porosity.

The processing parameters were further modified to reduce the phase size of the TiB_2 and improve densification. The commercial TiB_2 powder was initially ball milled alone for particle size reduction. Boron and methanol were then added for slurry milling. The Al and Mg were added following vacuum drying and ball milled. The resultant powders were hot pressed at 1500 °C. The hot pressed compact exhibited both ultra-fine particle size and intermixing along with minimal porosity. A typical region is shown in Figure 2.1.1. There are also numerous inclusions of unmixed $AlMgB_{14}$ within the hot pressed compact, not shown in Figure 2.1.1, most of which were in the 20 micron size range. A higher magnification optical image is also shown in the figure, revealing that some of the $AlMgB_{14}$ particles are clustered. Both of these figures demonstrate a fine intermixing of phases as was initially desired.

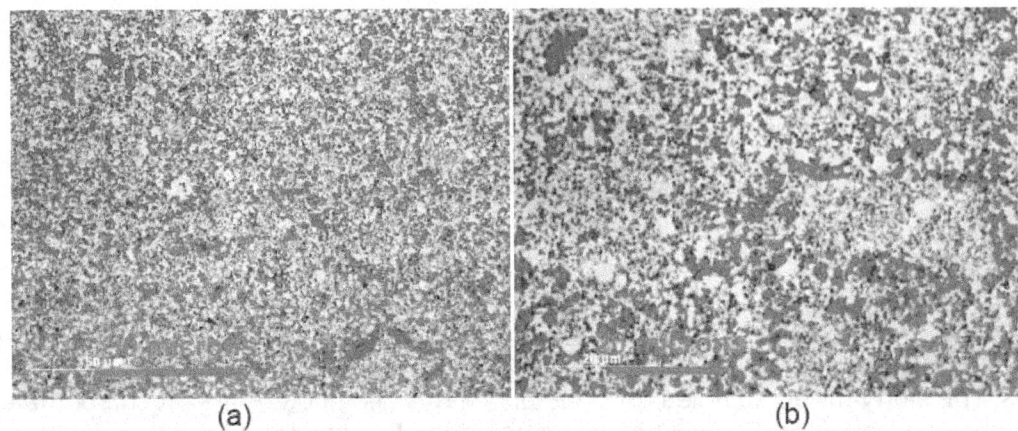

<table>
<tr><td>(a)</td><td>(b)</td></tr>
</table>

Figure 2.1.1. (a) Optical metallography of slurry milled $AlMgB_{14}/TiB_2$ composite. White phase is TiB_2, dark phase is $AlMgB_{14}$. (b) High magnification optical image of the sample at left. Clustering of the $AlMgB_{14}$ phase can be observed.

 Initial studies employed methanol as the slurry agent, which necessitated the milling of elemental B and TiB_2 prior to combining with Al and Mg.   A number of modifications to this approach to milling and mixing the powders were also examined. The milling periods were held to a relatively short duration in order to minimize the amount of Fe contamination from the media.   Samples were then consolidated by hot pressing at 1500ºC for 60 minutes. Optical micrographs of several representative samples are shown below in Figures 2.1.2 – 2.1.5 as examples of typical microstructures resulting from this processing.   It can be seen that the samples tend to be characterized by fine, intermixed $AlMgB_{14}$ (dark phase) and TiB_2 (light phase). A few larger particles of both phases are also observed. This would suggest that the Al and Mg are not breaking down completely during dry milling, leading to larger agglomerates of $AlMgB_{14}$ when the reaction sintering is complete.   It may also indicate that the B powder remains somewhat agglomerated following slurry milling and vacuum drying.

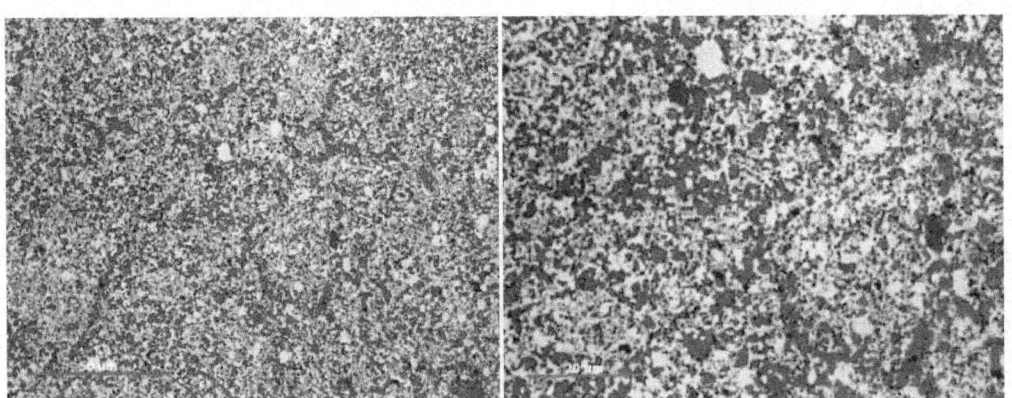

Figure 2.1.2. Premilled TiB_2 followed by methanol milling with B for 30 minutes. Vacuum dried and milled with Al and Mg 99 minutes. Scale bar left = 50 microns, scale bar right = 20 microns.

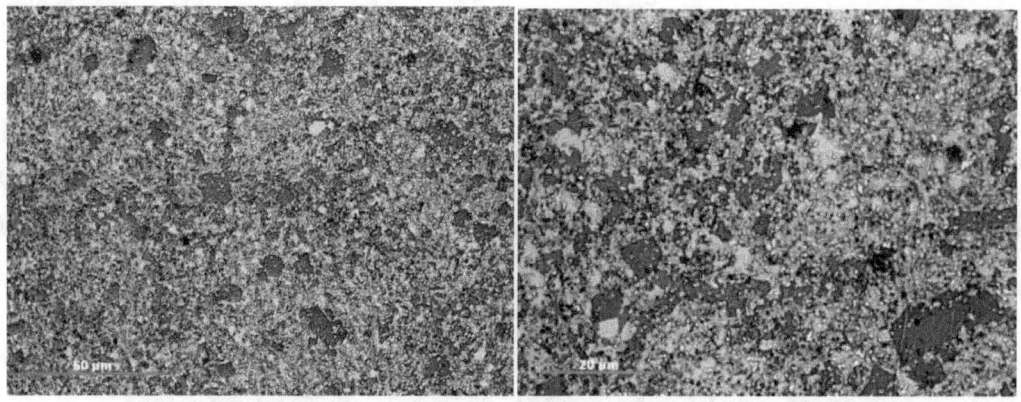

Figure 2.1.3. TiB$_2$ and B methanol milled 45 minutes. Vacuum dried and milled with Al for 5 minutes. Mg added and milled 99 minutes. Scale bar left = 50 microns, scale bar right = 20 microns.

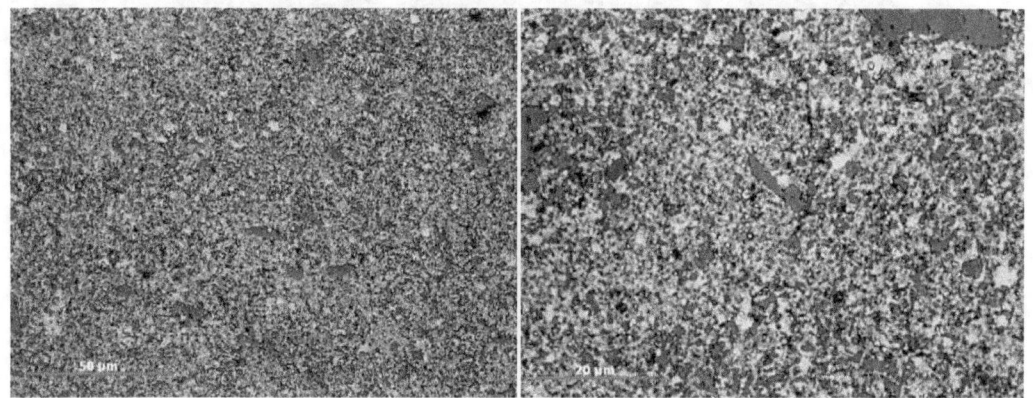

Figure 2.1.4. TiB$_2$ and B methanol milled with a single ¾ inch steel ball 10 minutes. Vacuum dried, added Al and Mg with five ½ inch balls and milled 60 minutes. Scale bar left = 50 microns, scale bar right = 20 microns.

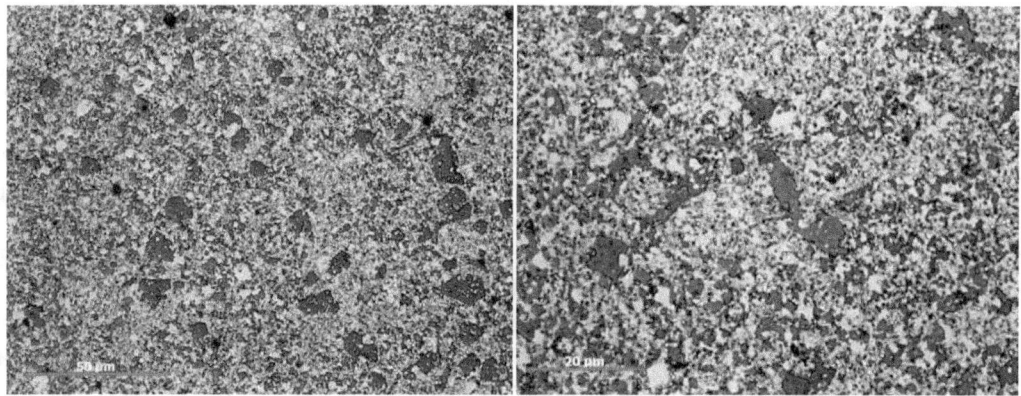

Figure 2.1.5. TiB$_2$ methanol milled 15 minutes followed by methanol milling with B for 15 minutes. Vacuum dried and milled with Al and Mg 99 minutes followed by 30 minutes of orbital milling. Scale bar left = 50 microns, scale bar right = 20 microns.

The goal of these various milling techniques is to reduce or eliminate the larger grains of $AlMgB_{14}$ that form during hot pressing. These regions probably nucleate at points enriched in either Al, Mg, or B that were not fully comminuted during ball milling. Figure 2.1.6 shows a region from the same sample as Figure 2.1.1. At lower magnification some features evocative of precipitation at grain boundaries can be seen. These regions may delineate the contact surfaces of agglomerated powder particles after mechanical milling. If the surfaces are Al or Mg rich, it seems likely that $AlMgB_{14}$ can preferentially form on these boundaries. Based on these results, different process control agents were employed to produce finer, uniform microstructures.

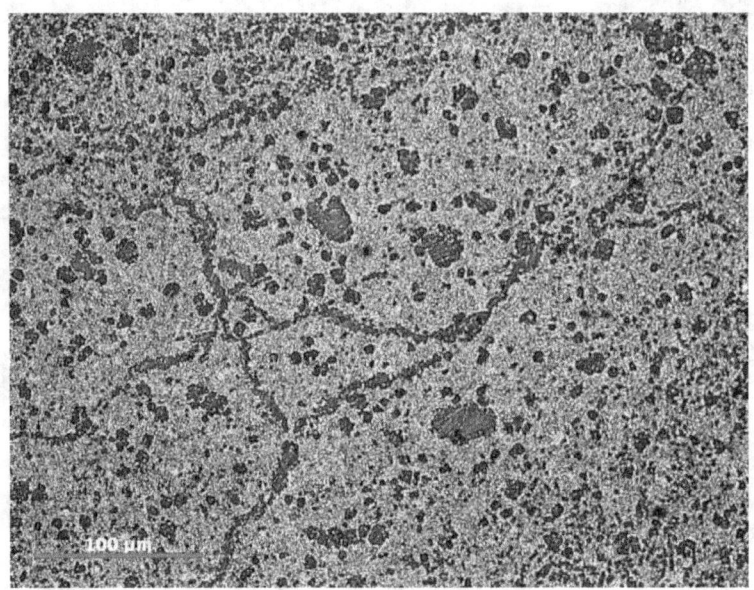

Figure 2.1.6. $AlMgB_{14}$ formation along apparent boundaries in hot pressed sample.

Scale bar = 100 microns.

Process Control Agents

Slurry milling of boride precursor powders has been demonstrated as a scalable method to reduce phase size and to mix the phases in a more uniform manner. Initial compacts prepared by this method have shown promising microstructures with minimal processing times compared to conventional dry processing. However, one issue that arises from methanol milling is a reaction between Mg and methanol that precludes formation of the $AlMgB_{14}$ phase. A new approach has been developed using stearic acid ($C_{18}H_{36}O_2$, or octadecanoic acid) as a process control agent. Borides prepared with stearic acid are also processed for comparatively short durations in order to reduce Fe contamination of the powder (i.e., less than 2 hours compared with 6 hours for conventional dry-milled powder). This is an important processing improvement in that metal inclusions resulting from the wear debris during comminution can range from a few percent up to as much as 15 to 20 percent in some samples. In order to reduce the amount of metal phase contamination and also to reduce the energy input required to process the material, we have investigated the use of stearic acid during processing. In other studies, we have found that incorporation of a small amount of stearic acid reduces particle agglomeration, and acting as a surfactant, enables the particle size distribution to reach near-nanoscale levels with minimal energy input. In order to reduce the energy intensity of processing in laboratory-scale $AlMgB_{14}$-TiB_2, samples were prepared with 1% stearic acid (by weight) and subjected to high-energy milling for 2 hours, as opposed to the conventional 12-hour milling time applied to previous samples. Flowability of the resultant powder was observed to be excellent, and the hot pressing produced compacts with erosion rate values comparable to the best baseline materials produced by 12-hour milling (0.5 mm^3/kg) . A secondary benefit is reduced iron contamination corresponding to the shorter milling time. This approach was carried over to the large-scale attritor milling processing, and several hundred-gram quantities of baseline $AlMgB_{14}$ powder were prepared using reduced impeller speeds. A multi-step processing program consisting of rotational speeds ranging from 300 to 500 rpm was employed instead of the earlier 1400 rpm speed. Analysis of the powder by x-ray diffraction showed that the level of iron contamination was significantly reduced. This was established by the absence of FeB peaks, which are the characteristic signature of Fe contamination in $AlMgB_{14}$. Samples of the powder hot pressed at 1400°C did not densify as completely as conventional high-energy milled powder, which is another indication that the Fe content is low. Iron has been shown to act as a sintering aid for both $AlMgB_{14}$ and TiB_2. When the powder was pressed at 1500°C, the resulting compacts were found to reach full density. Erosion tests were performed on these compacts, and the results compared very favorably with standard baseline $AlMgB_{14}$. Specifically, a steady state erosion rate of 1.75 mm^3/kg was observed, compared with 2.70 mm^3/kg for baseline high-energy milled $AlMgB_{14}$ and 2.05 mm^3/kg for the same material prepared with low-oxygen boron. (Recall that these erosion rate values correspond to baseline $AlMgB_{14}$ without TiB_2 additions.) This demonstrates that reduction in the amount of iron (or tungsten) contamination is highly beneficial to improving wear resistance. Figure 2.1.7 shows the dry erosion rates of stearic acid milled compacts compared to a similar standard 12-hour, high-energy milled samples. A trend is seen in which increased erosion resistance is obtained with increased milling time. Wet (slurry) abrasion tests using #400 grit diamond for the 0.5-, 3-, and 12-hour milled samples are in general agreement with the erosion data as shown in Figure 2.1.8.

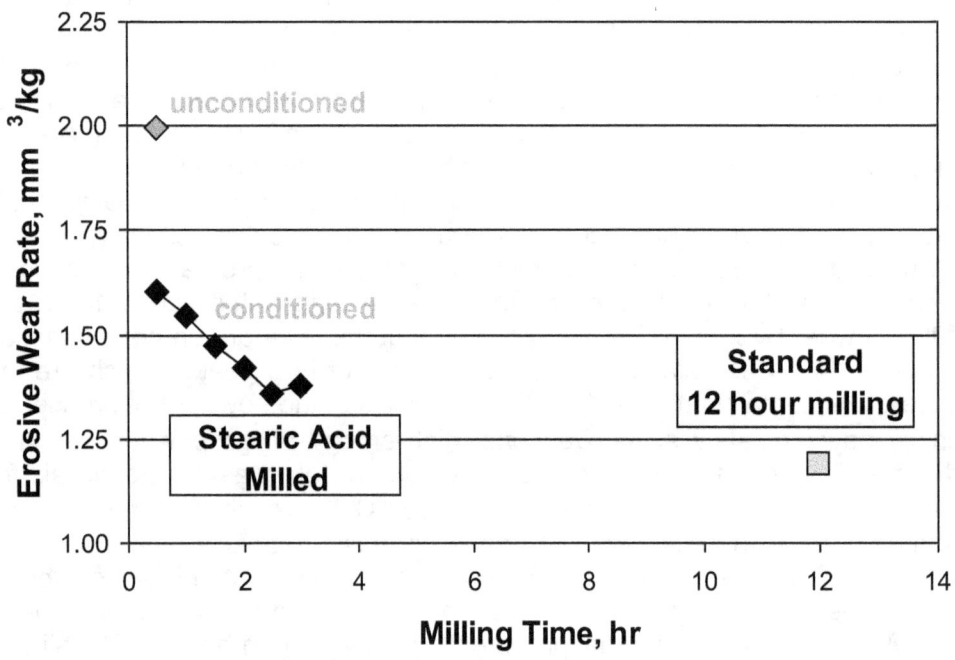

Figure 2.1.7. Erosive wear rates for stearic acid-milled $AlMgB_{14}$-70wt%TiB_2 along with a comparable standard.

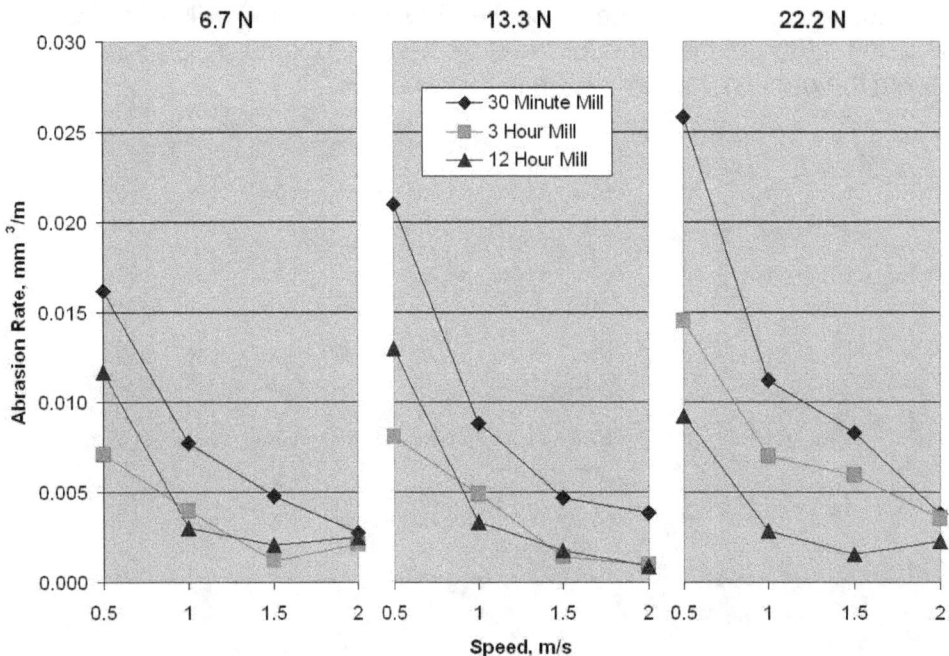

Figure 2.1.8. Wet abrasion data for stearic acid milled (0.5 and 3 hour) and standard high-energy milled (12 hour) samples.

Figure 2.1.9 shows the differences in microstructure between 2 and 2.5 hours of milling with stearic acid. There appears to be a significant change in morphology after a relatively short incremental increase in milling time. The microstructure on the right shows what appears to be growth of the boron-rich (dark) phase. This is supported by the somewhat faceted nature of some of these grains. Past "in-situ" samples where milling of TiB_2 and $AlMgB_{14}$ occurred simultaneously for longer times exhibited similar microstructures with faceted grains. XRD of current and past "in-situ" samples also show reduced $AlMgB_{14}$ peaks, indicating that this accelerated grain growth may reflect the formation of a new phase. The onset of the appearance of these faceted grains also coincides with the plateau or possible increase in erosion rate with further milling time.

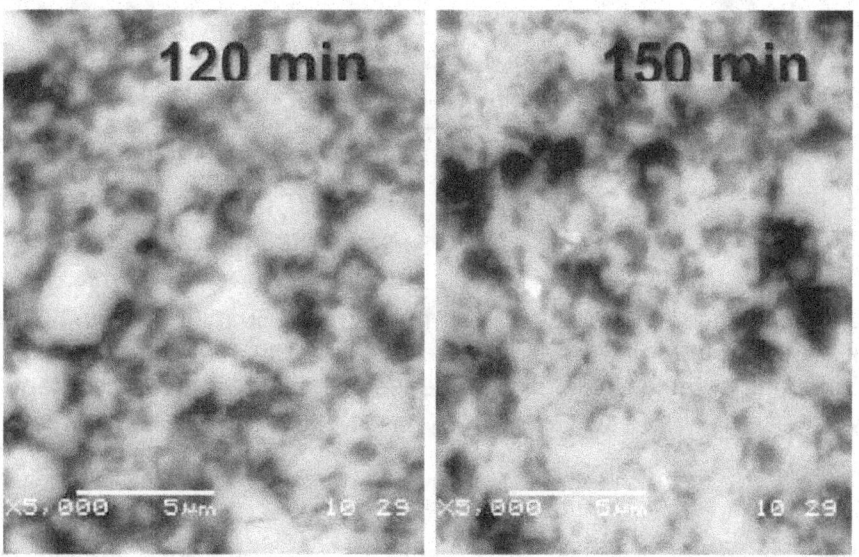

Figure 2.1.9. A comparison of stearic acid milled $AlMgB_{14}$-70wt%TiB_2 microstructures. (Left: 120 minutes; Right: 150 minutes.)

Short-duration vibratory milling with stearic acid $AlMgB_{14}$/TiB_2 composites ranging from 30 to 70 weight % TiB_2 was examined by x-ray diffraction. Samples of elemental Al, Mg, and B powders along with TiB_2 and 1 weight percent stearic acid were milled 60 minutes and hot pressed at 1500°C for 60 minutes. X-ray diffraction results shown in Figure 2.1.10 demonstrate the presence of $AlMgB_{14}$ in all samples.

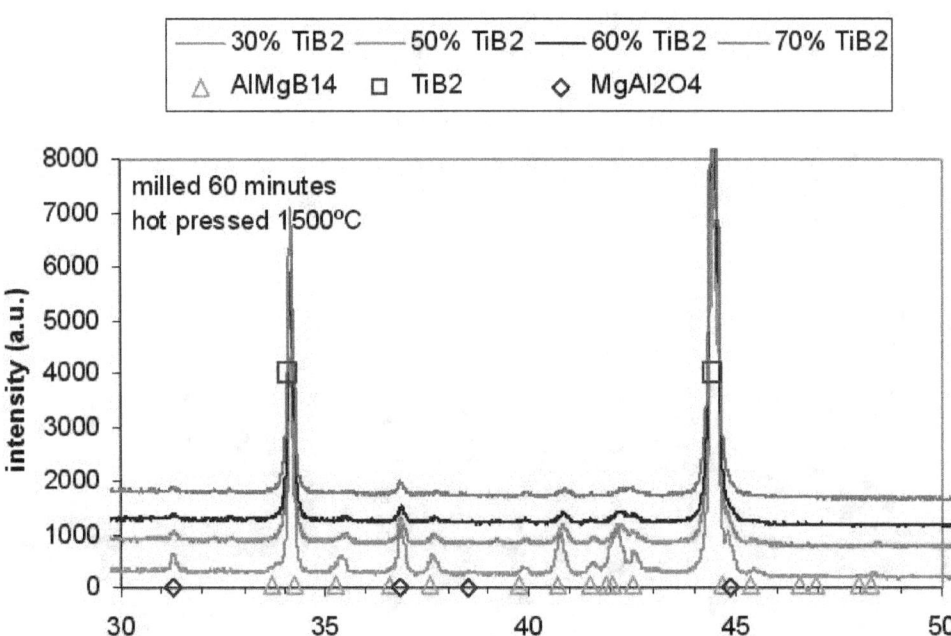

Figure 2.1.10. X-ray diffraction patterns of samples milled 60 minutes with 30, 50, 60, and 70 weight % TiB_2 (bottom to top). All samples were milled with 1 wt% stearic acid. Intensity of lines attributable to $AlMgB_{14}$ decrease with increasing TiB_2 content.

However, the reduced milling intensity also results in less iron contamination, which is both a benefit and a complication. Less iron results in a harder material without the deleterious FeB secondary phase, but Fe also acts as a sintering aid, facilitating densification of the boride powders. As shorter milling times also result in lower Fe concentrations, subsequent samples have been produced with controlled additions of Fe in order to maximize sinterability while minimizing energy intensity during processing and also the subsequent consolidation temperature. Much of this analysis is based on wear data, as erosion tests have previously been highly effective at identifying the balance between porosity and grain growth. Samples of $AlMgB_{14}+60$ wol. % TiB_2 were synthesized by milling the elemental constituents with stearic acid for times ranging from 30 to 180 minutes. The powder were then hot pressed for 1 hour at 1400°C and the wear resistance of each compact was evaluated by ASTM dry erosion testing. The samples were then subjected to abrasive waterjet erosion. The material loss was measured by profilometry and converted to an erosion rate based on total volume loss and the feed rate of the garnet abrasive. Figure 2.1.11 summarizes the results, showing that the correlation between waterjet and dry erosion testing is excellent. Further process control during waterjet erosion is expected to further improve the correlation. For instance, nozzle-to-sample distance and proximity to (porous) sample edge were not as consistent as desired during this first (proof of concept) test. Multiple profilometry measurements on each sample will also reduce statistical error.

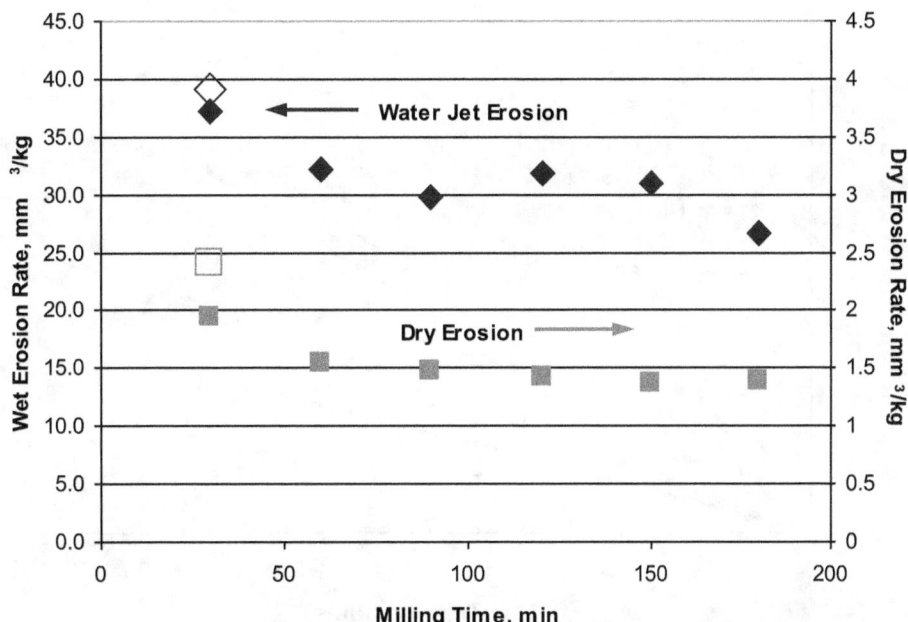

Figure 2.1.11. Comparison of erosion rates based on laboratory-scale ASTM dry erosion and from industrial waterjet machining of TB60-st samples milled with stearic acid. Open figures represent samples from unconditioned milling.

Figure 2.1.12 compares the dry erosion resistance of compacts prepared with process control agents and some of the dry-processed boride composites. The compacts prepared with process control agents are seen to exhibit a steady state mass loss comparable to the best dry-processed $AlMgB_{14}$-TiB_2 composite borides.

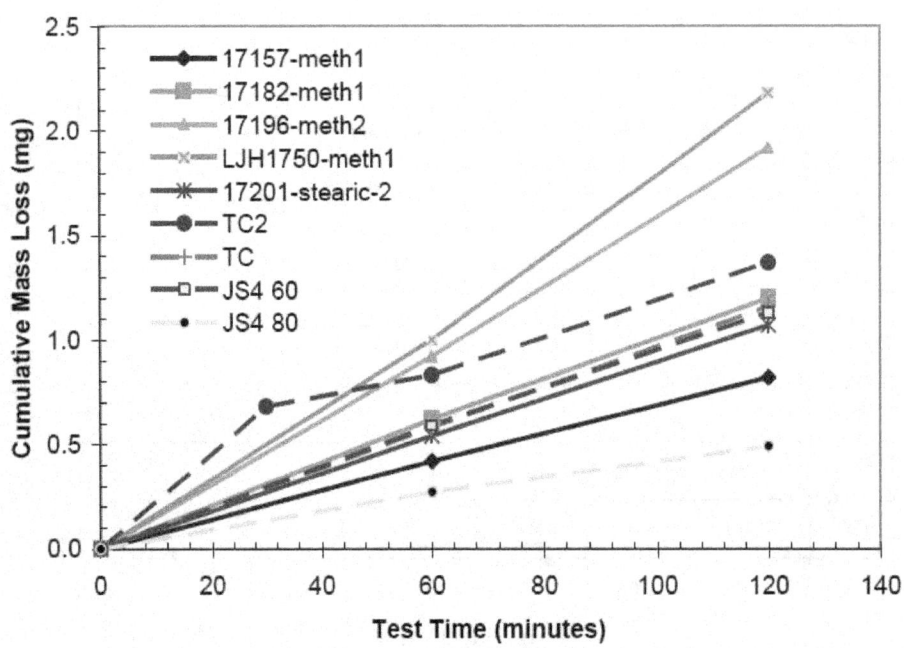

Figure 2.1.12. Erosion test results for samples prepared with milling aid (methanol or stearic acid) and short duration milling periods. All samples were hot pressed at 1500°C/60 minutes. Note the comparatively low mass loss for the composite prepared with stearic acid (17201 stearic-2). Refer to Table 2.1.1 below for sample descriptions.

Table 2.1.1. Sample descriptions for Figure 2.1.12. All compositions are $AlMgB_{14}$ + 70 wt. % TiB_2.

ID	Sample Description
1750-meth1	planetary milled baseline + TiB_2 + methanol, mechanically milled 30 minutes.
1796-meth2	B + TiB_2 + methanol, mechanically milled 30 minutes. Al and Mg were added and then MM 120 minutes
17157-meth1	B + TiB_2 + methanol mechanically milled 10 minutes. Add Al, Mg, MM 60 minutes
17182-meth1	TiB_2 + methanol, MM 15 minutes. Add B + methanol, MM 30 minutes. Add Al + Mg, MM 99 minutes.
17201-stearic-2	1 wt% stearic acid. MM 99 minutes.
TC	Best dry-milled boride compact from 2006.
JS4 60	Best dry-milled 60 wt. % TiB_2 composite prepared in 2008
JS4 80	Best dry-milled 80 wt. % TiB_2 composite prepared in 2008

These compacts were all examined in an SEM and the microstructures suggest that smaller phase sizes lead to improved erosion resistance. The sample prepared with stearic acid was found to possess the most refined microstructure. These samples demonstrate that short processing cycles are capable of producing compacts of comparable quality to the best dry-milled materials. The reduction in phase size, the fine interspersing of phases, and the reduction in porosity of the compact all appear to be key elements in the production of wear resistant materials.

The significance of this finding is that high quality wear resistant articles can be produced by industrially-scalable process technology with relatively short milling times. The process control agent does not appear to have a significant effect on erosion resistance.

In Situ Self Comminution

Alternative forms of milling were investigated as means of scale-up of processing to industrial quantities. The two major objectives include a reduction in Fe contamination (≤ 3 wt% or less as determined previously) and to facilitate scale-up of powder production. Most of the best materials are produced by high-energy vibratory milling, which is not directly scalable to large-volume production. First, a technique developed referred to as "In-Situ Self Comminution," or ISSC, was investigated, wherein boride powder in a sealed vial was vibratory mixed without the use of metallic media. The hard, abrasive nature of the powder is exploited in order to comminute, or induce size reduction, of the contents. In addition, the absence of metallic media offers the advantage of avoiding iron contamination, most of which arises from wear between the media and vial. The theory is that high-speed collisions between brittle TiB_2 and $AlMgB_{14}$ grains should induce transgranular fracture without the necessity of media. After only 2 hours of ISSC, fracture was not observed in the plate-like TiB_2 grains, although the grain edges have been noticeably rounded, as seen in Figure 2.1.13. The image on the right also shows $AlMgB_{14}$ particles which are not present in the left image. The $AlMgB_{14}$ particles were of irregular shape at the initiation of milling and further size reduction was not apparent. Size reduction due to collisions between particles has been observed in other configurations, such as impinging streams of erodent, but in this case the particles do not appear to achieve similar relative velocities to immediately induce fracture. It should be noted that some low energy milling processes require 50 to 100 hours for the desired processes to reach completion.

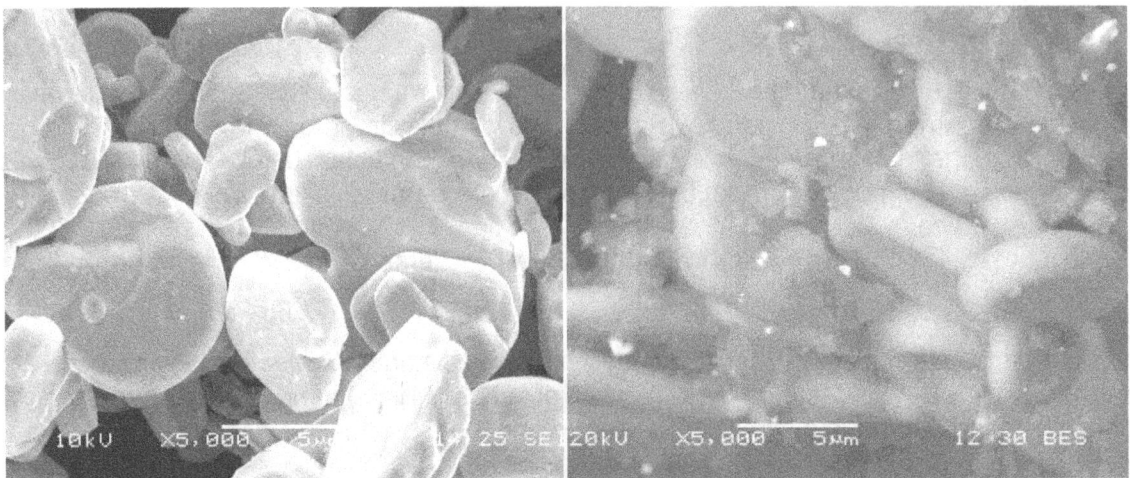

Figure 2.1.13: As received TiB_2 crystalline powder (left) and TiB_2 particles (with $AlMgB_{14}$) after 2 hours of in-situ self comminution (ISSC) (right). Note the rounding of the TiB_2 edges that has occurred as a result of the processing.

High Energy Ball Milling/WC Media

One of the pervasive obstacles to achieving unprecedented erosion and abrasion resistance in $AlMgB_{14}$-TiB_2 nanocomposites has been the presence of wear debris arising from mechanical comminution, most notably iron originating from the use of steel-based reaction vessels and media. Alternative milling approaches include using ceramic media instead of steel to reduce Fe impurities (sufficient amounts of Fe to aid sintering can be added to the initial mill mixture). WC media is the first candidate, as WC additions to $AlMgB_{14}$-TiB_2 composites were determined to be beneficial to hardness and toughness by Bodkin, et al. In the modified procedure, a single WC article was employed, based on previous studies that have shown multiple brittle media in a high energy vial tend to wear and fracture excessively. The resulting compact, following consolidation of the loose powder by hot-pressing at 1400°C, was observed to possess a lower density than typical baseline compacts. As shown in Figure 2.1.14, this is believed to be a consequence of the reduction in Fe impurities, consistent with our earlier findings that a small amount of metallic phase is needed as a sintering aid. In addition, it appears the extent of powder comminution was reduced with the use of fewer media; the grain size (as estimated by the spacing of spinel impurities) is larger in the WC milled material. Longer milling time (with the same media) should improve comminution. The Fe, which acts as a sintering aid, can be included as an intentional additive during synthesis in appropriate quantities, to achieve a higher compaction density. Previous studies estimate this to be ~3 wt% Fe, much lower than 10-15 wt% observed in current baseline samples.

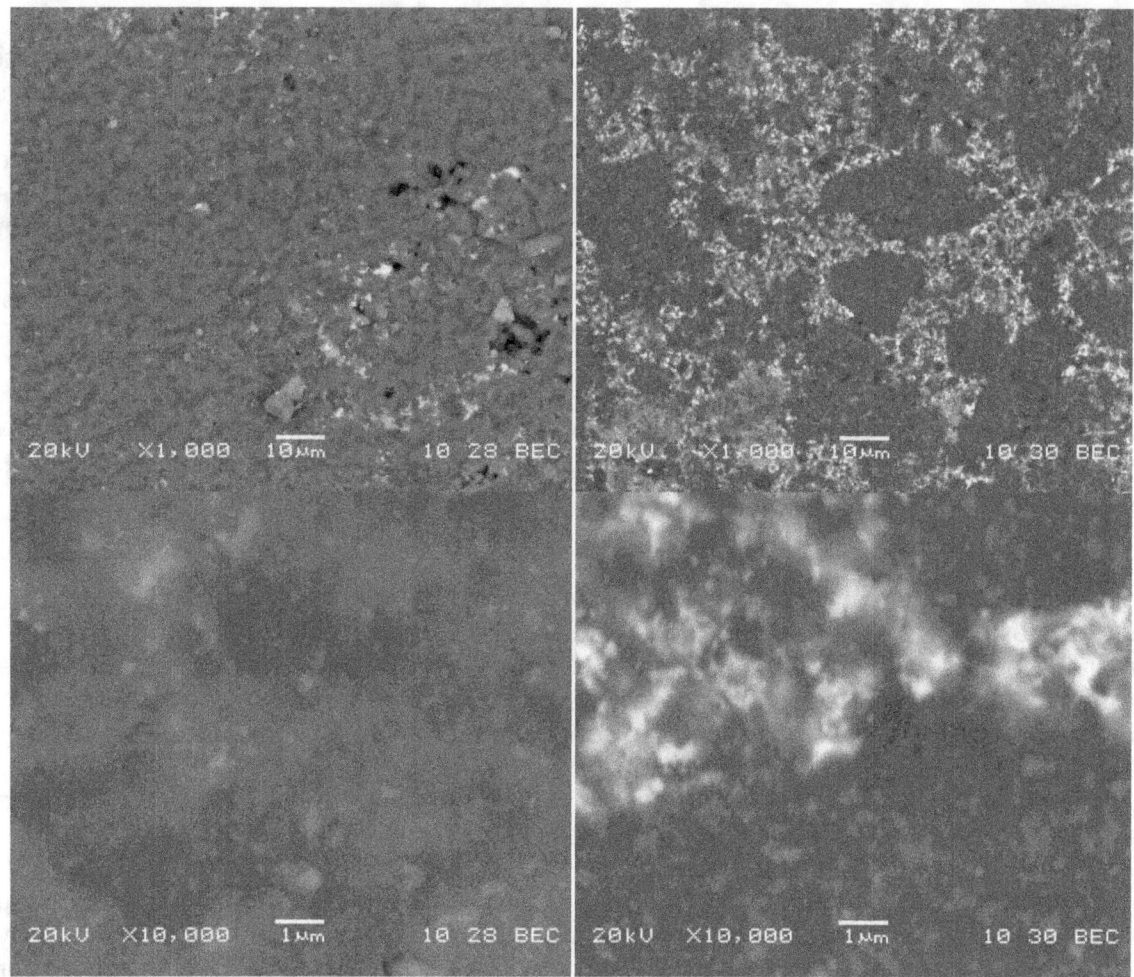

Figure 2.1.14: Baseline $AlMgB_{14}$ compact obtained from powder milled with WC media (left) compared to typical $AlMgB_{14}$ material milled with steel media (right). Bright regions indicate phases containing Fe and/or W. In lower images, grey spots indicate O (spinel) impurities, which roughly correlate with $AlMgB_{14}$ grain size.

Low Energy Attritor Milling

Powder prepared by the attritor mills has typically involved high impeller rotational speed to simulate the high kinetic energy attained in the vibratory laboratory-scale milling systems (HEBM). However, this procedure introduces high levels of Fe contamination into the powder, as well as accelerates impeller wear in the attritor. Preliminary studies at reduced energy levels in the attritor indicates that improved comminution and mixing of large scale powder batches compared to planetary milling can be achieved, while maintaining an acceptably low amount of iron contamination. The novel application of low speed rather than high speed attritor milling to form composite powders has been demonstrated on the Ames Laboratory attritor by [Ott 08]

Low-energy attritor milling was investigated as a means of reducing the energy intensity of processing large quantities of boride powder for sintering and/or consolidation. Conventional high-energy attritor milling results in significant Fe pick-up, however, by reducing the milling speed, the Fe contamination level was reduced to an extent that the characteristic FeB peaks in the x-ray diffraction pattern of hot pressed compacts were no longer visible. While attritor milling is capable of producing kilogram quantities of powder, fine metal powder presents a flammability hazard, which increases demands on handling and packaging requirements for shipment. By sintering the powders at temperatures ranging from ~800-1200°C, AlMgB$_{14}$ can be formed, with the resultant powder less reactive than the high surface area constituent metal and boron powders. Residues on furnace components after sintering have been associated with Mg and Al loss by vaporization during the high temperature treatment. While the nominal 1:1:14 stoichiometry of the starting powders is metal-rich compared with the Al$_{0.75}$Mg$_{0.78}$B$_{14}$ phase, this may not be the optimum ratio for the highest yield of AlMgB$_{14}$. MgAl$_2$O$_4$ remains and formation of boron-rich phases such as AlB$_{12}$ is also possible. It may be possible to ameliorate losses of Mg and Al to the spinel phase by a suitable adjustment to the nominal composition. Such a study would be beneficial in determining if there exists an optimum range of compositions between Al$_{0.75}$Mg$_{0.78}$B$_{14}$ and Al$_{1.0}$Mg$_{1.0}$B$_{14}$ that would enable reduced-temperature sintering while minimizing spinel content.

Intermetallic Precursors
Wear resistant boride materials based on AlMgB$_{14}$ are typically prepared from a mixture of the elemental precursors. Unfortunately, these elements are all subject to oxidation in air and quickly assume an oxide coating. During the course of alloy synthesis, these oxides act as diffusion barriers to alloy formation and also contribute to the formation of MgAl$_2$O$_4$ spinel, which is highly detrimental to the wear resistance of boride-based composites. Consequently, development of an alternate synthesis technique, bypassing the need for elemental powders, would offer a significant benefit in terms of processing cost and quality of the final product.

The first attempt to investigate the feasibility of employing intermetallic compounds as starting materials involved commercial grade AlB$_{12}$ and MgB$_2$. The two compounds were alloyed together in the appropriate ratios by high energy ball milling and the powder was subsequently hot pressed. X-ray diffraction of the resulting compact established that the desired AlMgB$_{14}$ phase was indeed formed during hot pressing. This is encouraging since lower cost intermetallic precursors would be expected to reduce processing costs in industrial applications. During the next quarter, we expect to perform additional studies to establish optimal processing conditions and investigate the possibility of employing other precursor compounds.

2.2. Consolidation

The consolidation of large powder batches is required to transition from laboratory scale to industrial scale articles. Numerous challenges exist in this arena. Powders must be fully consolidated to high density in order to maintain effective hardness and wear properties. The microstructure must be uniform and fine grain size must be maintained. Likewise, composition must be uniform throughout a consolidated compact. Consolidation must also be performed in an economical manner with readily available compaction technologies and techniques. And finally, the handling of powder prior to and during compaction must be accomplished in a safe manner. This section examines a number of techniques utilized to produce large compacts. Typically a slow progression from laboratory scale to large scale was followed and is documented for comparison. Emphasis was also placed on pre-reacting powders and their subsequent handling in order to improve the safety of steps leading to consolidation.

2.2.1 CIP/SINTER

One of the particularly attractive features of this project is the leveraging of materials synthesis and processing expertise of the National Laboratories with the scale-up, commercialization, and field-testing expertise of the industrial partners. As a starting point in this process, the evaluation of non-mechanical alloying processing techniques were investigated and weighed against the results of the existing database of hardness, abrasive, and erosive wear results on the laboratory-scale samples. Two samples were prepared utilizing elemental Al, Mg, and B powders blended in a vibratory mixer for 180 minutes. One sample, designated JH-16-151A, utilized B powder vacuum outgassed at 1400°C. The B for the second sample, JH-16-154A, was vacuum outgassed at 1100°C. The vibratory mixed powder was cold isostatically pressed (CIP) at 60 ksi to form a rigid green body. A portion of each green body was sintered under flowing Ar at 1400°C/60 minutes. Each sample lost some Mg due to outgassing, however the x-ray diffraction patterns of the samples, shown in Figure 2.2.1, indicate only lines indexed to $AlMgB_{14}$ and $MgAl_2O_4$.

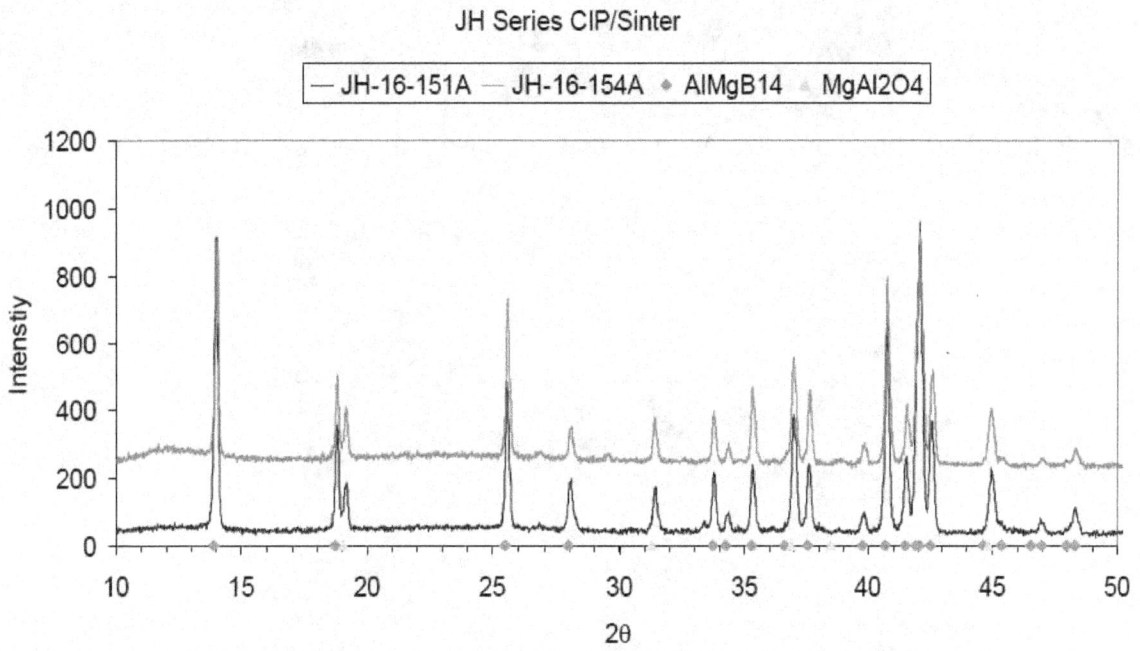

Figure 2.2.1. X-ray diffraction pattern of $AlMgB_{14}$ formed from CIPed/sintered blended elemental powders.

These two samples clearly demonstrate that a simple, economical vibratory blending of standard commercial powders can be used to prepare fully alloyed material for subsequent consolidation. This result is highly significant in that it shows that the desired hard compound can be formed by a sintering operation at ambient pressure. This provides a more commercially friendly approach to production of $AlMgB_{14}$. Further processing can be extended to the formation via blending, CIPing, and sintering of an $AlMgB_{14}$-70 weight percent TiB_2 composite.

2.2.2 Hot Isostatic Pressing

Numerous powder samples have been prepared by pilot-plant-scale processing methods such as planetary and attritor milling. Quantities of these powders have been hot isostatically pressed (HIPed) at 1200°C and 310 MPa for 60 minutes in stainless steel vessels. Figure 2.2.2 shows the optical micrographs for HIPed samples of various genres of powder. Figure 2.2.2 (a) shows a baseline AlMgB$_{14}$ prepared by vibratory powder blending of the elemental constituents, Figure 2.2.2 (b) shows baseline AlMgB$_{14}$ powder prepared by planetary milling, and Figure 2.2.2 (c) is AlMgB$_{14}$ + 30 weight percent TiB$_2$ prepared by high speed attritor milling.

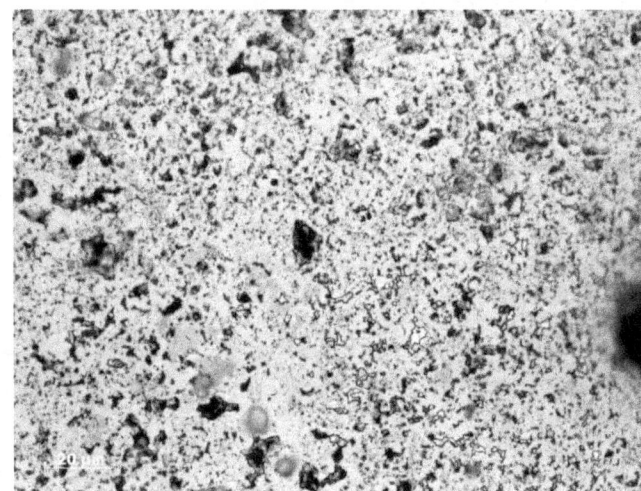

Figure 2.2.2 (a). Blended baseline AlMgB$_{14}$ powder HIPed at 1200°C/310 MPa/60 minutes.

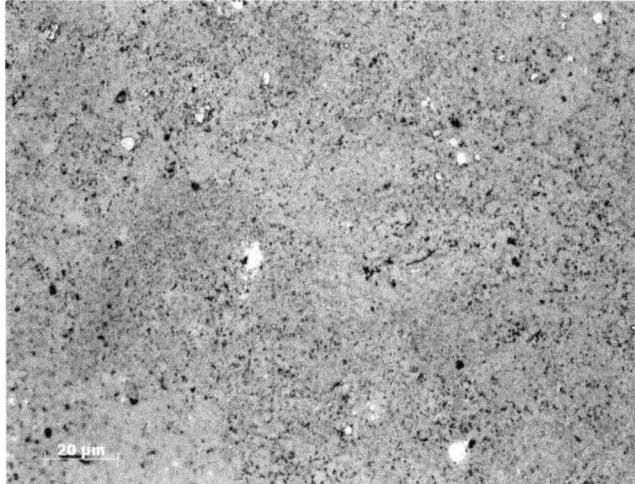

Figure 2.2.2 (b). Planetary milled baseline AlMgB$_{14}$ powder HIPed at 1200°C/310 MPa/60 minutes.

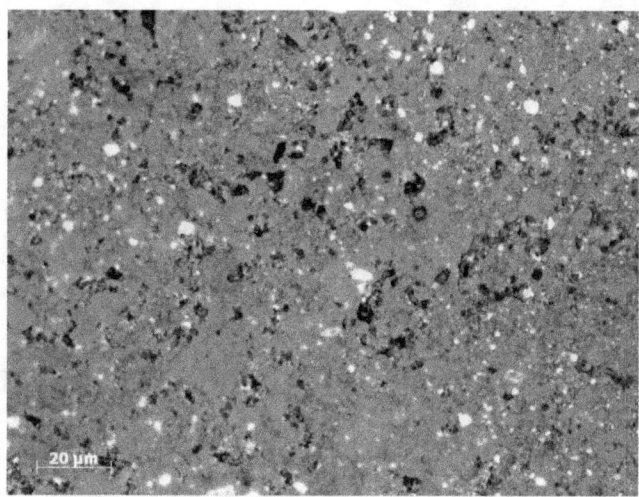

Figure 2.2.2 (c). Attritor milled $AlMgB_{14}$ + 30 weight percent TiB_2 powder HIPed at 1200°C/310 MPa/60 minutes.

It can be seen that the blended powder contains a high level of open porosity. The attritor and planetary milled powder also contains some porosity, but both densified to a degree compatible with subsequent containerless HIPing at higher temperatures to complete the densification process. It has been observed that planetary milled and blended powder samples exhibit higher porosity than vibratory milled powder under the standard hot pressing parameters of 1400°C/103 MPa for 60 minutes. Lower applied pressure for longer consolidation times appears to reduce porosity. These results are encouraging for high pressure Dynaforging, discussed below, at temperatures closer to 1400°C, utilizing short compaction times to achieve high densities while maintaining fine microstructures.

2.2.3 CIP and Variations on Hot Consolidation

Two approaches have been identified as possible commercial-scale processing routes for $AlMgB_{14}$ and other wear-resistant ceramic materials:

- Hot consolidation of blended elemental powder
- Hot consolidation of pre-reacted powders

Both techniques offer a distinct advantage over the mechanical alloying/hot pressing method initially applied to the preparation of laboratory scale research samples. High-energy mechanical milling (HEM) typically results in Fe contamination in the milled powders from wear debris during processing with steel media. In contrast, the production of pre-reacted $AlMgB_{14}$ powders with reduced contamination has been accomplished either through blending the elemental powders, followed by cold isostatic pressing (CIP) and sintering, or by a combination of pre-densification by CIP followed by sintering the milled powders. The $AlMgB_{14}$ phase, along with $MgAl_2O_4$ spinel, has been shown to form at temperatures ranging from 1000°C to 1400°C.

Due to constraints on maximum applied pressure in many commercial-scale hot presses, the primary focus has involved compensating for a lower applied pressure by increasing the temperature in order to reduce porosity in the finished compact. Two tests were performed to assess the feasibility of this approach; one to examine microstructure resulting from a low applied pressure hot pressing and the second to examine the high temperature stability of a hot isostatically pressed $AlMgB_{14}/TiB_2$ composite.

The first test utilized pre-reacted powders formed by blending the elemental powder constituents, followed by CIP/sintering to form $AlMgB_{14}$ and grinding to form powder. The powders were hot pressed at 1400°C and only 6 ksi for 60 minutes, rather than the typical applied pressure of 15 ksi. The product was a structurally sound, albeit coarse porous compact. The x-ray diffraction patterns of both the sintered pre-reacted powder and the subsequent hot pressed sample are shown in Figure 2.2.3.1. The sintered compact exhibits peaks traceable to $AlMgB_{14}$ and spinel. This demonstrates that compacts can be formed by hot pressing pre-reacted powder but that high density compacts require higher applied pressure when pressed at 1400°C.

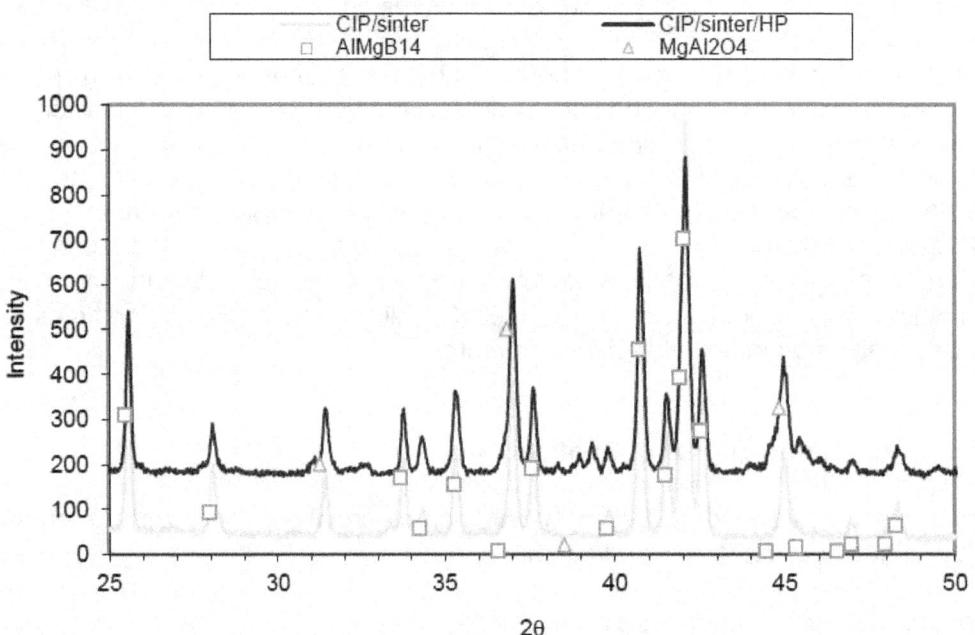

Figure 2.2.3.1. X-ray diffraction pattern of CIP/sintered material (bottom) and the same material followed by grinding and hot pressing (top).

The thermal stability of $AlMgB_{14}$ at 1700°C was determined by sealing a sample of hot isostatically pressed material containing 50 weight percent TiB_2 in Ta and heat treating 60 minutes at 1700°C. The x-ray diffraction pattern after hot pressing revealed only $AlMgB_{14}$, spinel, and TiB_2, indicating that hot pressing the pre-reacted powders at 1700°C is feasible.

Samples of mixed phase boride powder prepared at Ames were hot-pressed by Greenleaf Corp. to determine if higher temperatures combined with lower pressure could produce comparable material to the lab-scale samples. If successful, this would enable the pressing of larger samples in presses with lower pressure capability. Table 2.2.3.1 compares typical temperature and pressure used at Ames Laboratory with those used at Greenleaf.

Table 2.2.3.1: Temperature and pressure settings during hot-pressing and resulting densities

	Temp. (°C)	Pressure (MPa)	Density (g/cm^3)
Ames	1400	106	3.26
Greenleaf	1500	34	3.02
Greenleaf	1600	34	2.97
Greenleaf	1700	37	2.96
Greenleaf	1800	34	2.92

As seen in Figure 2.2.3.2, density tends to decrease with increasing pressing temperature. The largest incremental decrease, from 1400 to 1500°C, can be largely attributed to the substantially higher pressure applied at 1400°C. The overall downward trend is likely due to the application of pressure after the initial phase of sintering and necking has begun. In all samples, maximum pressure was applied at or near the onset of maximum temperature. In a previous study a similar trend was seen when pressing baseline AlMgB$_{14}$ from 1300 to 1500°C where maximum density was achieved at 1300°C. Again, pressure was applied upon reaching the pressing temperature and it was concluded that necking of particles before reaching maximum temperature reduced the effectiveness of pressure on consolidation. Maximum hardness and mechanical properties were still seen after pressing from 1400-1450°C perhaps due to better particle bonding despite higher porosity. This would suggest that pressure should be applied at 1300°C or lower before reaching peak temperature.

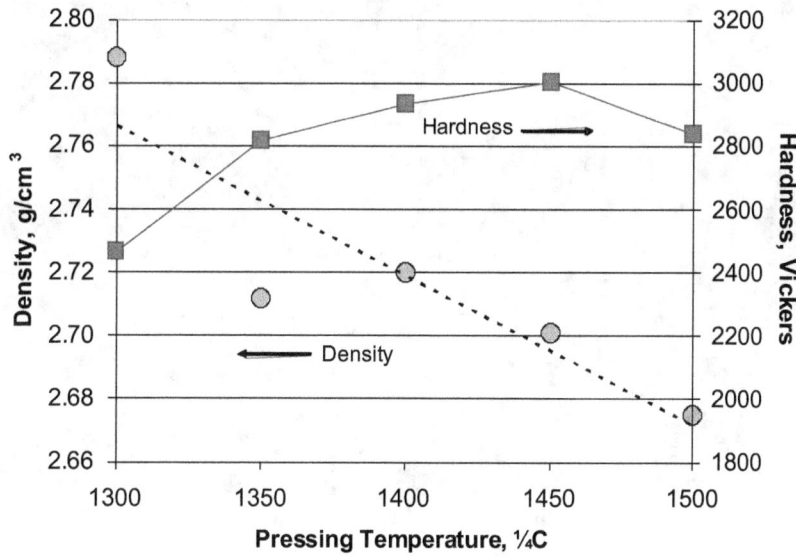

Figure 2.2.3.2: Hardness and density of baseline large-scale AlMgB$_{14}$ vs pressing temperature for samples hot pressed at Ames Laboratory and Greenleaf Corp.

Electron microscopy shows little difference in microstructure between samples pressed at 1400, 1500, and 1600°C. Figure 2.2.3.3 shows the first significant change after pressing at 1700°C. Grain size has not increased significantly, but the majority of sub-micron grains present in lower temperature samples have disappeared. Since the sub-micron microstructure is considered to be a significant contributor to the composites' high hardness and erosion resistance, this would be expected to result in a reduction of mechanical properties for samples pressed above 1700°C. At 1800°C, most fine grains are eliminated, and there has been some grain growth.

Higher porosity and larger pores can be seen in the 1500°C sample in Figure 2.2.3.4, which correlates with their respective measured densities.

The 1800°C sample exhibited the highest erosion rate as seen in Figure 2.2.3.5, followed by the 1700°C sample, which agrees with predictions based on their respective microstructures. Erosion rates were lowest for the 1600°C sample, which does not follow the expected trend with respect to density. This may be within the margin of error, or may be due to other factors revealed by microscopy of the erosion craters. Erosion of the Greenleaf samples revealed pre-existing cracks within the compacts, which affects erosion resistance. Due to the random arrangement of the flaws, direct comparison of the erosion rates of the samples becomes difficult due to the higher error.

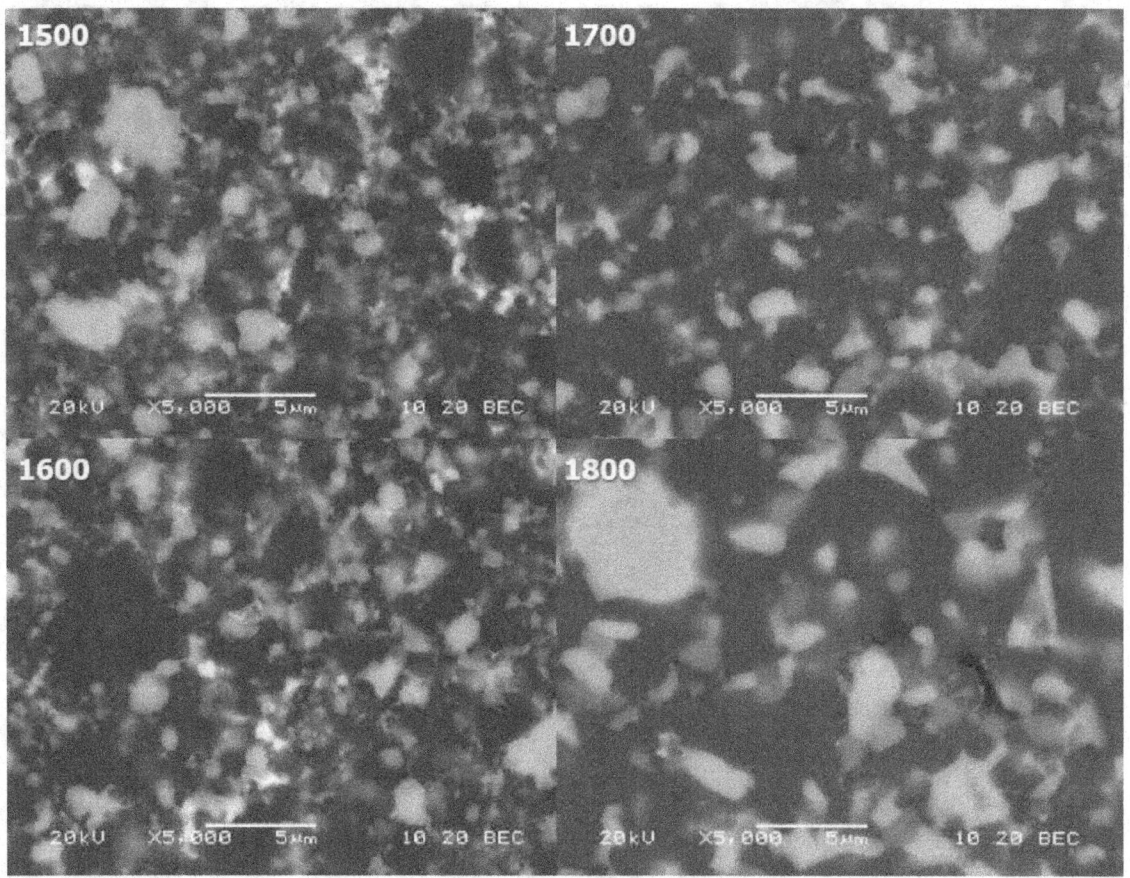

Figure 2.2.3.3: Backscattered micrographs of samples listed in Table 1, with light TiB_2 and dark $AlMgB_{14}$ grains.

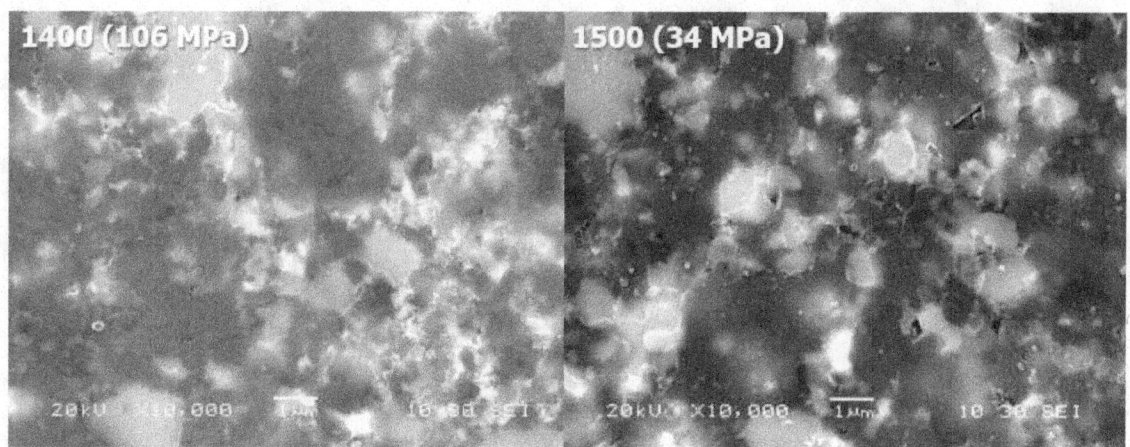

Figure 2.2.3.4: Secondary micrographs of samples pressed at 106 and 34 MPa with some visible porosity.

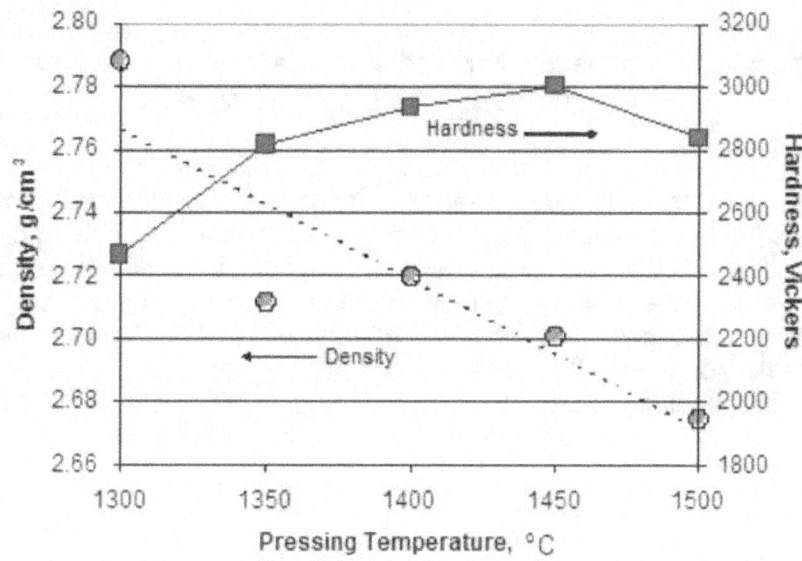

Figure 2.2.3.5: Erosive losses for the samples listed in Table 2.2.3.1.

Results of these studies indicate that the use of pre-reacted powders is viable for high-temperature, low-pressure hot consolidation. However, alternate consolidation paths are also suggested. Since the formation of the $AlMgB_{14}$ phase begins at 900°C and the reaction is completed by 1100°C, these temperatures indicate that HIPing can be attempted in conventional, low-cost stainless steel vessels under an inert atmosphere. The use of stainless steel is preferable to the expensive Ta used for high-temperature HIP vessels. Alternately, reacted powder could be formed by sintering in sealed stainless steel to obtain powder for hot consolidation. To test the compatibility of steel as a reaction vessel, blended baseline powder was loaded into a stainless steel crucible and sealed with a slight positive He pressure. The crucible was heated to 1200°C for 60 minutes. There appeared to be no reaction between the steel crucible and the powder. An x-ray diffraction pattern of the heat-treated powder, shown in Figure 2.2.3.6, shows the typical pattern obtained from reacted powder, namely $AlMgB_{14}$ and $MgAl_2O_4$ spinel.

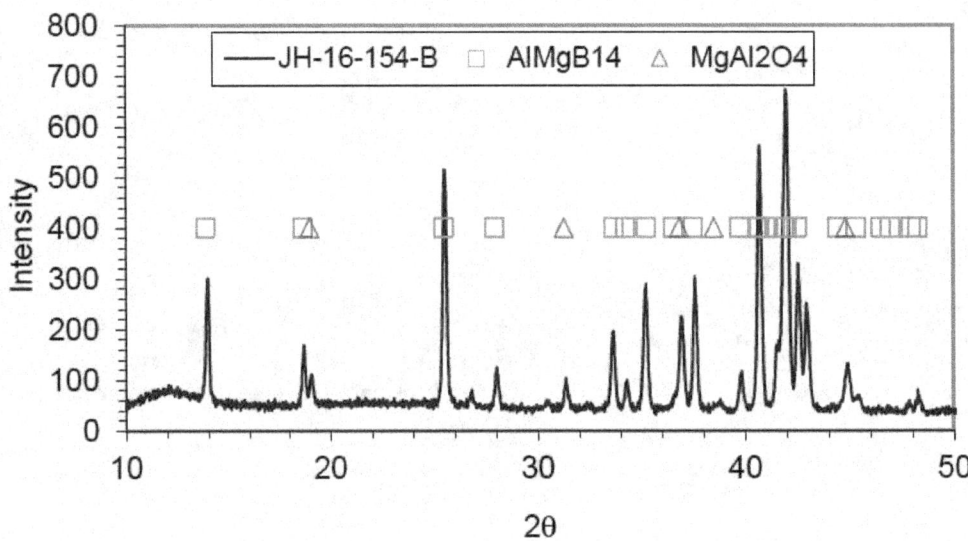

Figure 2.2.3.6. X-ray diffraction pattern of mixed elemental powder sintered in sealed stainless steel crucible at 1200°C for 60 minutes.

These laboratory-scale studies provide guidance for future commercial-scale processing of advanced wear-resistant ceramic materials. In addition to establishing time and temperature conditions for densification of the material, these studies can be applied to specific commercial-scale densification approaches likely to result in the highest density. One example is the rapid densification Dynaforge technology developed by Carpenter Powder Products, Inc., which combines rapid heating of an encapsulated sample with application of high pressure (hundreds of MPa). Likewise, commercial scale hot presses utilizing larger dies can also match production to the above parameters.

2.2.4. Dynaforge Rapid Consolidation

One of the limiting processing issues related to scale-up and commercialization of laboratory-based materials is the ability to densify powders into net shaped articles in a cost-effective manner while retaining the attractive properties observed in the research-scale samples. Carpenter Powder Products, Inc. is an active partner in this project and is developing advanced technology for consolidating nanopowder of $AlMgB_{14}$ and related compositions. Their Dynaforge process offers the potential for providing a pathway for commercial-scale consolidation of these materials. It should be noted that conventional hot pressing requires one hour to achieve acceptable densification at a temperature of 1400°C. Two batches of pilot-plant mixed-phase $AlMgB_{14}$-50 vol % TiB_2 powder were first cold isostatically pressed (CIPed) and then sealed under vacuum in pyrex tubes. The CIPed specimens fractured into multiple sections within the pyrex tubes during transit, but remained suitable for Dynaforging. These specimens were Dynaforged at 1204°C with an applied pressure of 120 ksi for 30 seconds and 60 seconds, shown respectively in Figure 2.2.4.1 and Figure 2.2.4.2 below.

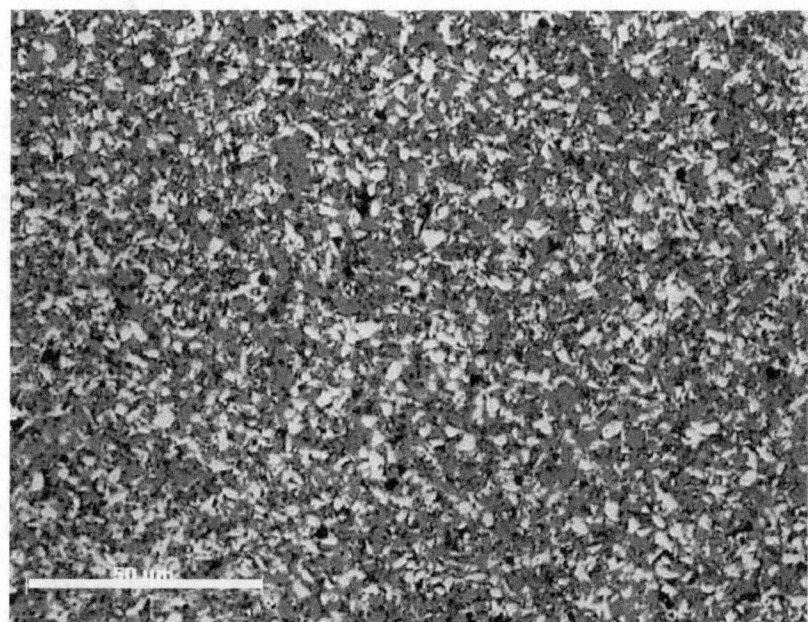

Figure 2.2.4.1. Optical micrograph of pilot plant AlMgB$_{14}$-50 vol % TiB$_2$ powder Dynaforged at 1204°C for 30 seconds. Light phase = TiB$_2$, gray phase = AlMgB$_{14}$, and black is porosity. Scale bar = 50 microns.

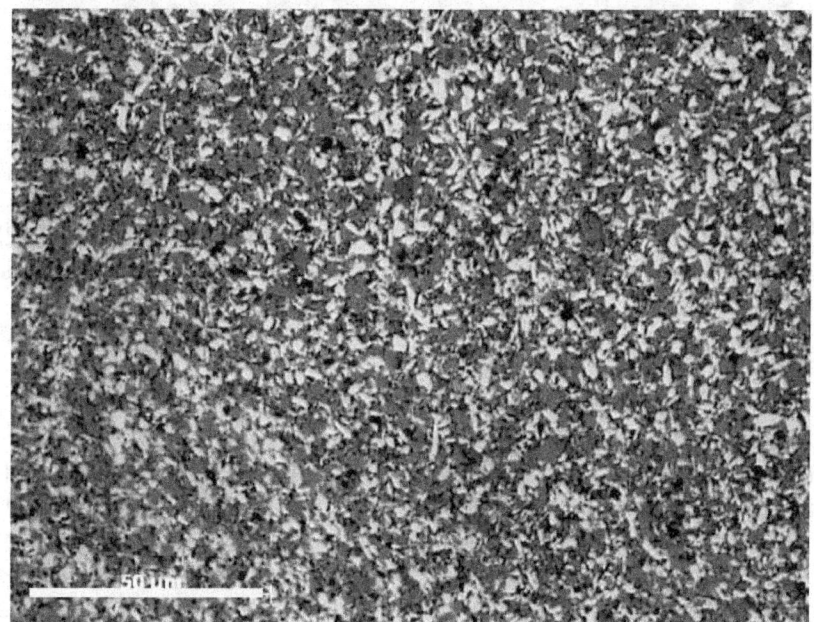

Figure 2.2.4.2. Optical micrograph of pilot plant AlMgB$_{14}$-50 vol % TiB$_2$ powder Dynaforged at 1204°C for 60 seconds. Light phase = TiB$_2$, dark gray phase = AlMgB$_{14}$, and black is porosity. Scale bar = 50 microns.

Although porosity remains evident in these specimens, the degree of densification at the low sintering temperature of 1204°C is highly encouraging. The micrographs indicate that Dynaforging at higher temperatures should enable full densification for the pilot plant pre-alloyed powder in 30 to 60 seconds.

Both of these samples possessed a dense but thin outer layer and a porous inner volume, indicating that higher temperatures are required for full densification. While awaiting completion of the furnace upgrade, a third sample was Dynaforged under identical temperature and pressure for 120 seconds. The increase in duration did appear to extend the region of high density further inward compared with the previous two samples, but full density at the center was not yet achieved, as seen in Figure 2.2.4.3.

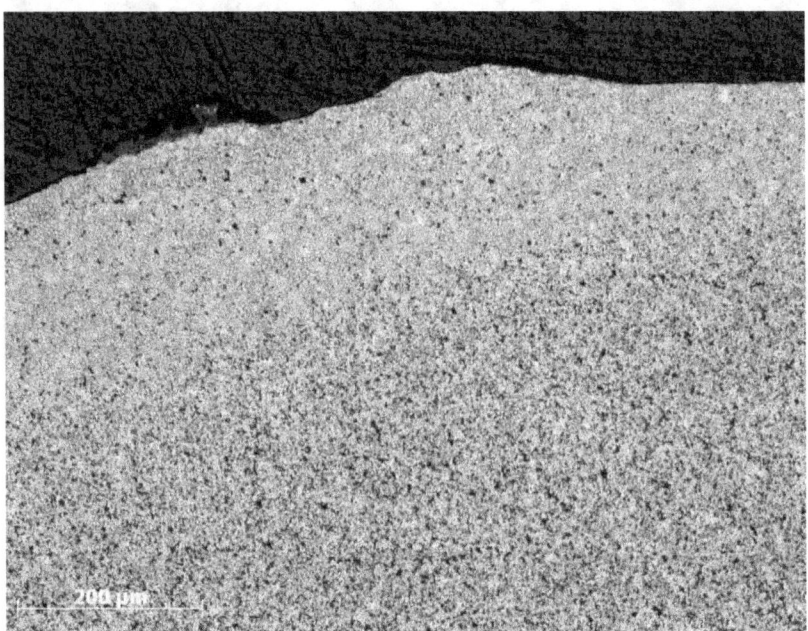

Figure 2.2.4.3. AlMgB$_{14}$-70 wt% TiB$_2$ Dynaforged at 1204°C/120 ksi/120 seconds. The sample exhibits a dense outer shell and porous inner core.

An upgrade of the Dynaforge furnace and press to achieve a temperature of 1425°C and a maximum pressure of 1.1 GPa was not completed by the end of this project. It is expected that this upgrade would significantly improve densification throughout the compact.

2.2.5. Consolidation with Co-Mn binder (Dynaforge and hot press)

AlMgB$_{14}$-based nanocomposites have primarily been produced by powder metallurgical techniques involving mechanical alloying and hot pressing. This approach leads to the incorporation of Fe from wear debris into the processed powder. While the Fe forms an FeB impurity phase during hot consolidation, the iron also serves a beneficial role as a binder/sintering mechanism to aid in densification of the compact. Co-Mn has been developed as a high quality binder for AlMgB$_{14}$-based grit particles, but the use of this binder requires that the AlMgB$_{14}$ grit be relatively phase pure. To this end, AlMgB$_{14}$ formed via cold isostatic pressing (CIP) followed by high temperature sintering was employed in the initial binder phase trials. The sintered body consists of AlMgB$_{14}$ sans Fe impurity phases with low spinel second phase, but when hot pressed by itself possesses too low a density for use in bulk applications. However, when ground to a fine powder, this material should serve well in a compact along with the Co-Mn binder.

The initial phase of this study involved vibratory blending pre-reacted $AlMgB_{14}$ – 50 wt. % TiB_2 powder with 15 wt% gas atomized Co-Mn, as a binder phase. The powders were cold isostatically pressed to 60 ksi after blending and the resulting green body was sealed in a thick-walled pyrex tube. Consolidation at CPP was carried out at 1200°C and 120 ksi (827 MPa) for 10 seconds. The outer diameter of the compact was reduced from 10-10.6 mm in the as-CIPed condition to 8.9-9.4 mm following compaction. The length was also decreased, from 82.6 mm to 79.5 mm. Since the recovered mass was equal to the starting mass, no penetration of the glass into the sample occurred. An examination of the resulting compact showed that the outer layer reached a higher ultimate density than the center, as seen in Figure 2.2.5.1. Density gradients are not uncommon during consolidation and their presence indicates the need to optimize processing conditions such as time, temperature, and pressure. The effect of porosity on hardness is manifest by the difference in microhardness values; the outer regions of the compact reached a hardness of ~ 28GPa while the center was closer to 19 GPa. For a complex, multi-phased ceramic containing 15 wt. % binder and consolidated for only 10 seconds, these results are highly encouraging considering the fact that porosity was observed in all locations and that compaction occurred for only 10 seconds.

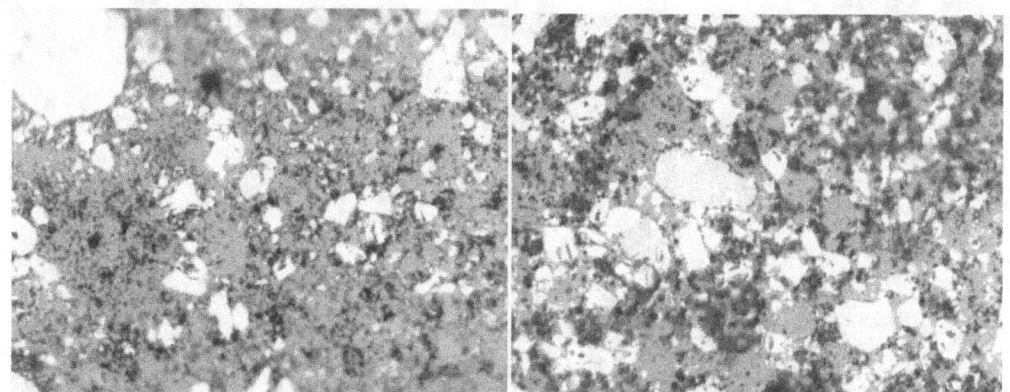

Figure 2.2.5.1. Micrograph showing the microstructure at the surface (left) and center (right) of $AlMgB_{14}$ – 50 wt. % TiB_2 powder with 15 wt% gas- atomized Co-Mn following Dynaforging. (1000x) Note the dark regions, indicating porosity, in the center-region micrograph.

Additional samples of $AlMgB_{14}$-50 wt. % TiB_2 with 15 wt.% gas atomized Co-Mn powder were Dynaforged in a sealed pyrex ampoule by Carpenter Powder Products at 1200°C and times ranging from 10 to 20 seconds. A maximum pressure of 827 MPa was applied to each sample. After removal from the ampoule, the consolidated cylindrical samples all exhibited a graded density, with the highest at the outer edge and progressively decreasing toward the center. The densification depth roughly followed the amount of time at temperature, with the shortest dwell time corresponding to the most shallow densification depth and the longest to the most extensive densification.

Erosion tests were performed on pieces recovered from the outer shell of the compacts. While these samples tended to be irregular in shape and not of the standard geometry, it was possible to determine mass loss as a function of time during erosion with 100 μm alumina grit and establish a qualitative comparison between the materials. As with the densification, a decrease in specific mass loss (mass loss normalized to initial mass) was observed with increasing consolidation times. The sample removed from the 20-second consolidation run exhibited essentially no mass loss after 30 minutes of erosion, compared with a 30% mass loss in the 10-second sample.

Since the Dynaforge furnace was not capable of attaining 1400C during the duration of this program, a control sample was prepared using baseline AlMgB$_{14}$ pre-reacted powder but consolidation was achieved by hot pressing at 1400C for 30 minutes. According to the binary phase diagram, the Co-Mn liquidus is ~ 1380°C. Upon melting of the Co-Mn phase, liquid state sintering is presumed to dominate over solid diffusion. Figure 2.2.5.2 shows an example of how the Co-Mn phase dramatically improves densification; porosity is evident in regions beyond the diffusion range of the Co-Mn binder.

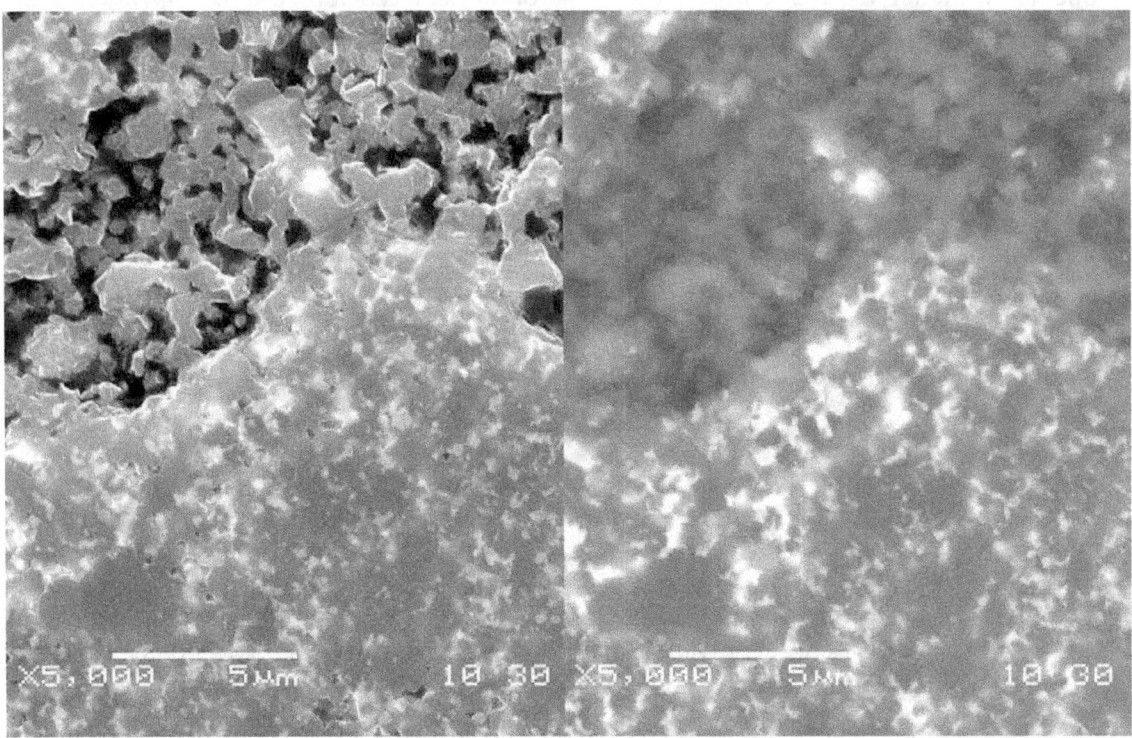

Figure 2.2.5.2: Topographical (left) and compositional (right) micrographs of the same region in hot-pressed AlMgB$_{14}$ containing Co-Mn, showing densification due to Co-Mn flow (bright). Some porosity remains in the upper portion of the figures where Co-Mn is not present.

Both the Dynaforge and hot press samples demonstrate that Co-Mn additions can aid in sintering, lowering pressure, temperature, and/or time required to densify the composites. The distribution of the Co-Mn powders was not optimal in these samples. As discussed previously, modifications to the Carpenter Dynaforge system were not finished by the completion of this project. However, the available results indicate that rapid consolidation combined with higher temperatures can result in high-density samples with well dispersed Co-Mn binder.

2.2.6. Spark Plasma Sintering

Oak Ridge National Laboratory examined the application of spark plasma sintering (SPS) to the $AlMgB_{14}$-based powders as an alternative method of hot consolidation. In SPS consolidation, heating of the sample is accomplished by a DC current through the sample itself, instead of from an external heating element as in conventional hot pressing. SPS offers the advantage of a much higher heating rate than conventional hot pressing (up to 1000K/min compared with 10K/min) which means that issues associated with extended soak times at high temperature, such as an undesirable coarsening through grain growth or a reaction between Co-Mn and $AlMgB_{14}$, become dramatically reduced.

Similar to a hot pressing furnace, an SPS furnace applies a compressive load to a powder sample in a die while simultaneously heating the material to high temperatures. The difference is that, while a hot press uses a set of heating elements inside the furnace chamber but external to the die, an SPS furnace runs current directly through the loading rams and the sample itself resulting in direct heating of the powder. In addition, it is theorized that the electric field in the system is concentrated at the contact points between powder particles and leads to intense localized heating and plasma formation within the powder compact, which may enhance sintering rate. The SPS furnace used for the consolidation experiments at ORNL is shown in Figure 2.2.6.1.

Figure 2.2.6.1. Thermal Technology spark plasma sintering furnace at Oak Ridge National Laboratory.

For the initial consolidation trials, 8.5 grams (2.35 cm^3 at final density) of a 50/50 $AlMgB_{14}$-TiB_2 composite powder mixture was loaded into a graphite die having an inner diameter of 1 inch and an outside diameter of 3 inches. A small hole was drilled into the side of the die to within 0.25 inches of the sample position for insertion of a thermocouple to measure the sample temperature during the sintering run. The die was wrapped with a blanket of insulating graphite felt before being loaded into the SPS furnace chamber. The felt reduces heat loss to the water-cooled metal chamber walls. Figure 2.2.6.2 shows the graphite die in position inside the furnace chamber.

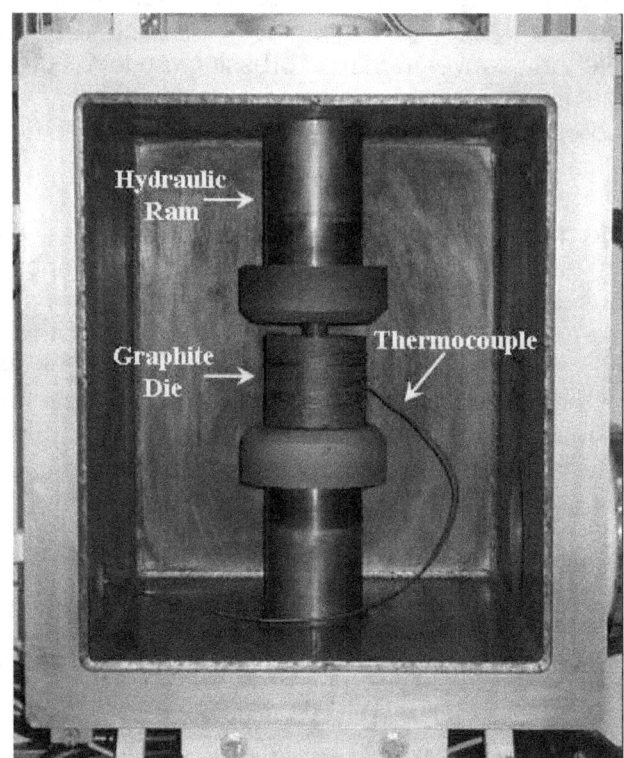

Figure 2.2.6.2. Furnace chamber of the SPS furnace with the graphite die in place.

Prior experiments at Ames Laboratory using a conventional hot press showed that a sample density of >95% theoretical could be achieved by sintering at 1500°C for 1 hour at 15,000 psi. Because of uncertainty regarding the graphite die strength and the unknown response of the $AlMgB_{14}/TiB_2$ material in the SPS furnace, the first trial was run at 1300°C and 3,000 psi. Measurement of the sample after the furnace run indicated a density of only 54%. For the second trial run, the operating conditions were modified by increasing the temperature to 1400°C and the pressure to 9,000 psi. Subsequent examination of the sample showed that the density had been increased to 74% of theoretical. Because of the large volume of material needed for this demonstration, powder material having the TB60 composition (40 vol.% $AlMgB_{14}$ / 60 vol.% TiB_2) was prepared at ORNL for the large consolidation experiments. SPS sintering tests of the powder in the small die showed lower density than that observed previously with Ames powder – 86% vs. 99+%. Similar behavior had previously been observed and can be attributed to the blending methods and atmosphere control of the Ames powder preparation process.

Two powder batches were prepared at Ames Laboratory for additional densification investigation. The first powder batch was an attritor-milled mixture of Al, Mg, and B having the baseline $AlMgB_{14}$ composition. The second powder batch was a TB60 composition utilizing the same powder with the addition of 60 vol % mechanically milled TiB_2. The attritor-milled powder has lower Fe content and thus reduced sinterability by conventional hot pressing. As shown below, this does not hinder the densification of the powder by SPS at the right settings.

Samples of the AlMgB$_{14}$ powder mixture were consolidated in a conventional hot pressing furnace to establish a baseline for comparison with subsequent SPS furnace runs. It was found that hot pressing at 1400°C and 15 ksi pressure for 60 minutes achieved a density of 95.3% of the theoretical value. Increasing the temperature to 1500°C and 15 ksi for 60 minutes improved the density to 97.7% of theoretical. Samples of the same powder consolidated in the SPS furnace reached 98.2% of theoretical density at 1400°C and 15 ksi for 30 minutes; a higher density at 100°C lower temperature than in the standard hot pressing furnace.

The density for the hot pressed AlMgB$_{14}$/TiB$_2$ mixture was 95.0% of theoretical when pressed at 1500°C and 15 ksi pressure for 60 minutes at Ames Laboratory. When consolidated in the SPS furnace at ORNL, it was found that the AlMgB$_{14}$/TiB$_2$ mixture had a density of 99.3% of theoretical when heated to 1400°C at 15 ksi pressure for 60 minutes. This is a significant increase in density at a lower temperature when compared to the hot-pressed material and may result in appreciable improvements in hardness, wear, and erosion properties for this material. A summary of the sample densities that were determined in these comparison tests is shown in Table 2.2.6.1. The densities were all measured by Archimedes method using ethanol.

Table 2.2.6.1. Percent of theoretical density of consolidated samples processed under spark plasma sintering (SPS) and conventional hot pressing (HP) at 15ksi. TB00 is AlMgB$_{14}$, TB60 is AlMgB$_{14}$-60 wt. % TiB$_2$ composite.

| | Pressing Conditions | | | | | | |
Material	SPS 1200°C 30 min	SPS 1300°C 60 min	SPS 1400°C 5 min	SPS 1400°C 30 min	SPS 1400°C 60 min	HP 1400°C 60 min	HP 1500°C 60 min
TB00	89.5%	--	--	98.2%	--	95.3%	97.7%
TB60	--	88.4%	95.2%	99.2%	99.3%	--	95.0%

Figure 2.2.6.3 summarizes graphically the time and temperature conditions for consolidation of some of the AlMgB$_{14}$-60 wt. % TiB$_2$ (TB60) and single-phase AlMgB$_{14}$ (TB00) powders in the SPS furnace, as well as a conventional hot press (HP). Lower pressing temperatures were insufficient to achieve reasonable density. The figure shows that SPS is superior to HP in that both shorter time and lower temperature can produce a denser article.

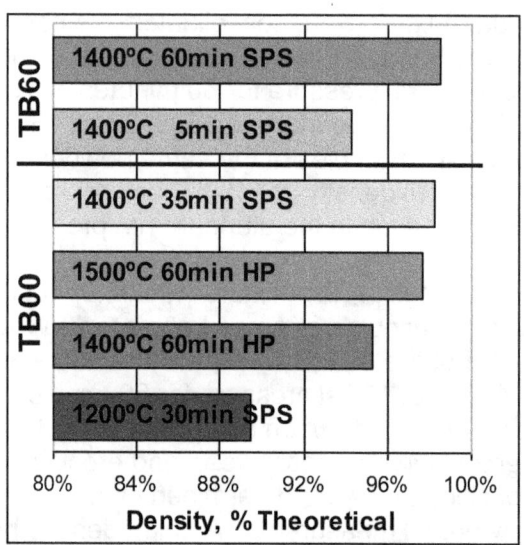

Figure 2.2.6.3. Density of AlMgB$_{14}$ (TB00) and AlMgB$_{14}$-60 wt. % TiB$_2$ (TB60) samples pressed at ORNL by both SPS and HP.

The microstructures of TB60 samples consolidated by SPS at 1400°C with varying times were evaluated. Figure 2.2.6.4 shows SEM micrographs of samples following 5 and 60 minutes pressing. The images show regions of high porosity after 5 minutes, and still higher magnification reveals the retention of fine grains in the same sample. Figure 2.2.6.3 shows that densification tends to become sluggish after 30 minutes; therefore it may be sufficient to terminate sintering at that time in order to retain as much of the fine microstructure as possible. With respect to erosion resistance, it may be that the benefit of a finer microstructure more than compensates for the slight difference in density; this has been observed before in series of 100% TiB$_2$ samples [Peters 02].

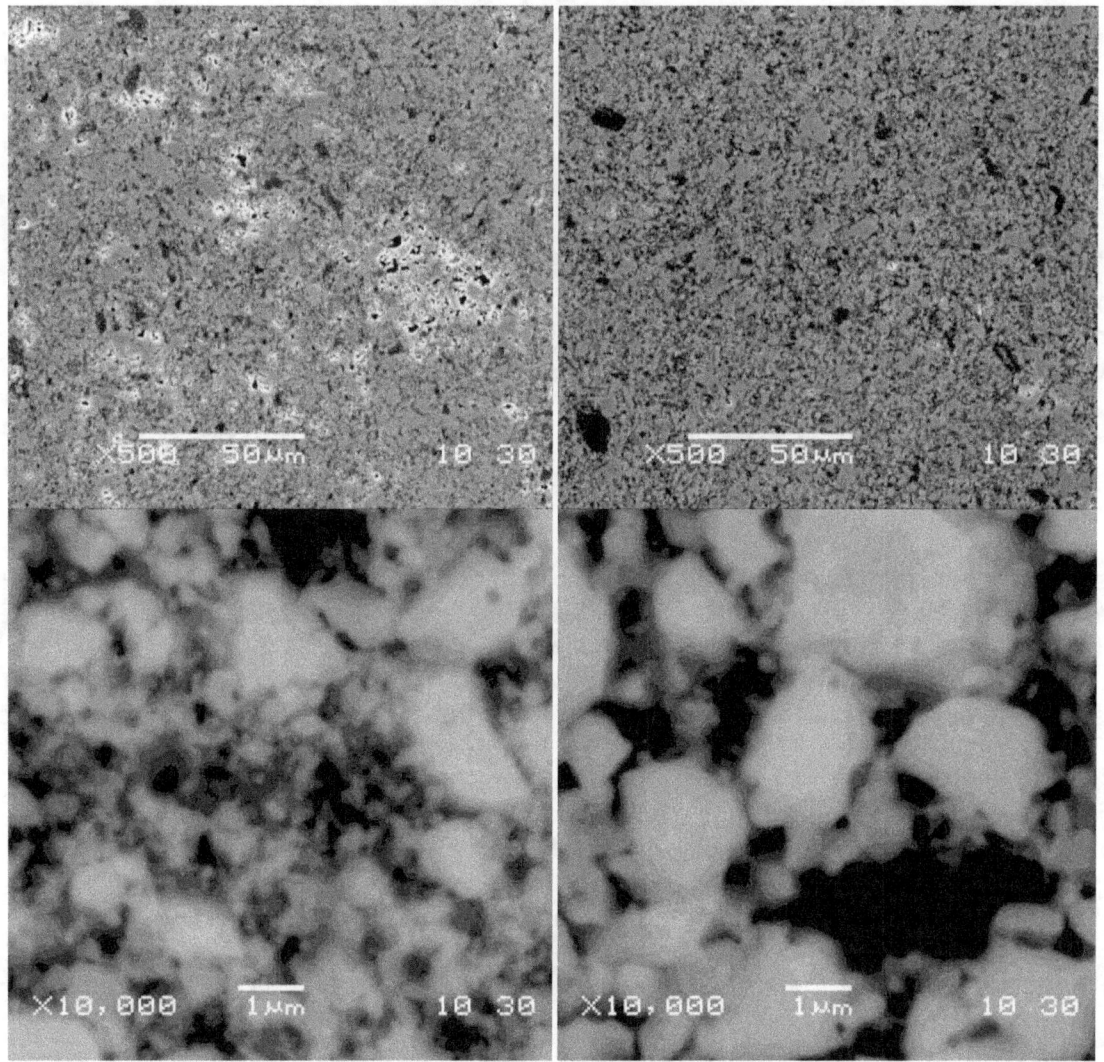

Figure 2.2.6.4: Microstructures of SPS TB60 samples pressed at 1400ºC for 5 minutes (left) and 60 minutes (right). Note the significant reduction in porosity with increased grain size with increasing time.

In addition to the samples listed in Figure 2.2.6.3, a TB60 mechanically alloyed powder blend was consolidated by SPS at 1300ºC for 60 min. As shown in the table, the lower temperature sample had a sintered density of only 88.4% compared to 99.3% for the sample run for the same time at 1400ºC. This suggests that 1400ºC is near the minimum temperature that can be used to obtain fully consolidated material of this composition in the SPS furnace. Typical microstructures for two of the samples are shown below; the black diamonds are Vickers microhardness indentations.

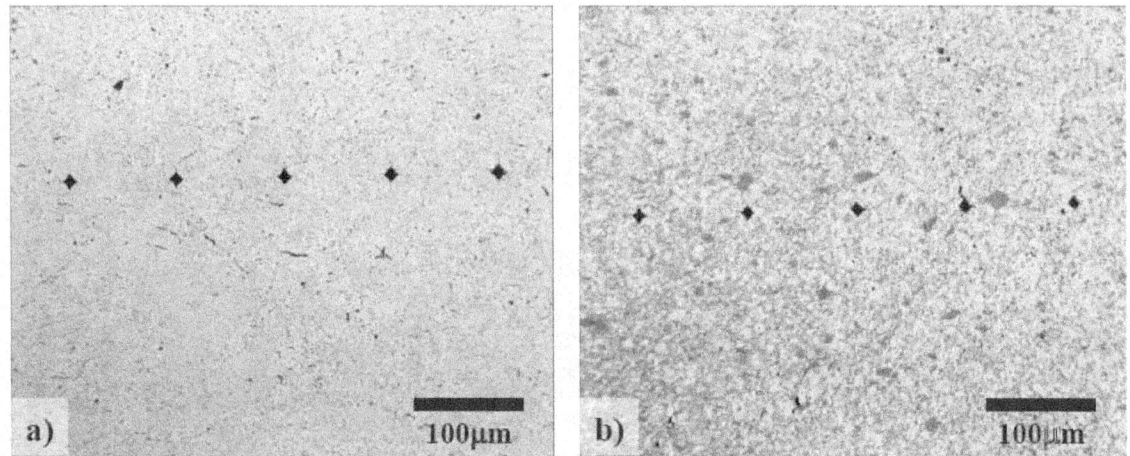

Figure 2.2.6.5. Optical micrographs of a) AlMgB$_{14}$ and b) AlMgB$_{14}$/TiB$_2$ consolidated by SPS showing Vickers indents and the uniform microstructure of the materials.

These tests with higher quality powder have shown that the direct heating process in the SPS furnace can reduce the temperature required for full consolidation of TB00 and TB60 samples by 100°C or more when compared to conventional uniaxial hot pressing. One of the goals of the project is the demonstration of a consolidation method for producing large samples (~50 cm^3) of boride ceramic. Initial SPS sintering tests were conducted using a die that produced 25.4 mm (1 inch) diameter (~2.2 cm^3) samples. A larger graphite die having the capacity for a 50.8 mm (2 inch) diameter sample was utilized in the SPS furnace to produce a compact weighing more than 185g consolidated at 1400°C for 30 min at 15 ksi pressure. This powder weight corresponds to a volume of greater than 50 cm^3 for a fully consolidated TB60 composition. Consequently, the SPS process was shown to be capable of handling the large size sample and producing a bulk volume of consolidated material. A large TB60 sample that was sintered in the SPS furnace is shown below, along with smaller samples produced in the 25.4 mm diameter die.

Figure 2.2.6.6. Photograph of TB60 samples showing a comparison of the size of 25.4 mm diameter (small) disks and the 185g, 50.8 mm diameter (large) sample produced by spark plasma sintering (SPS) during this reporting period.

2.2.7. Surface Preparation

Closely related to the characterization and optimization of these materials is the approach to surface preparation, which is critical for unambiguously establishing hardness and indentation toughness. From a production standpoint, finishing of consolidated material is needed to prepare the final component, such as, for example, a cutting tool. In the laboratory, polishing of these hard composites generated major delays and backlogs in the characterization. Hand grinding of the samples on bonded diamond wheels was once necessary to remove flashing and surface roughness from as-pressed compacts, requiring many hours per sample. Automated mechanical polishing relieves some of the human burden, with grinding of multiple samples in the same amount of time; however, some hand grinding was still needed to expose flat surfaces for the final polishing. More costly, consumable polishing supplies are also required during machine polishing. In an attempt to expedite the polishing process, a high-speed diamond grinder was used to produce parallel faces on the compacts. Due to the aggressive nature of the process, fracture was not uncommon and the risk of damaging unique specimens became too high. Wire-EDM was investigated as a way to obtain parallel faces on both sides of the compacts and in considerably less time. The high conductivity of the TiB_2 phase eases the process and compositions as low as 30 wt% TiB_2 can be machined in this fashion. The parallel faces of the samples require much less polishing and are ideal for ultrasonic, erosive wear, and conductivity studies.

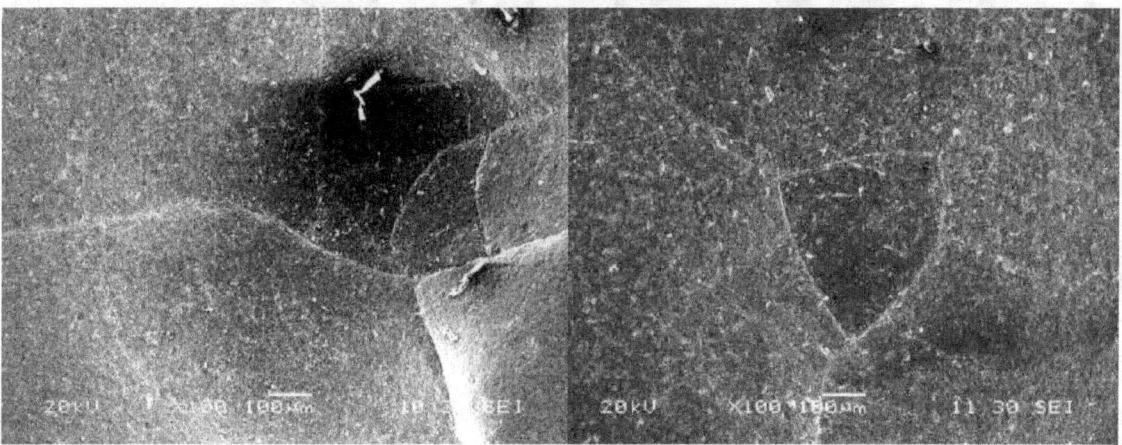

Figure 2.2.7.1. An example of pitting within erosion craters due to pre-existing flaws caused by aggressive finishing techniques. Samples pressed and finished at Greenleaf.

The results of this EDM trial are quite promising. High speed diamond grinding is the industry standard for shaping tooling materials. This method works well with WC-Co composites containing a ductile metal phase. Yet in the ceramic $AlMgB_{14}$-TiB_2 composites, this grinding technique was found to introduce subsurface fracture, as shown in the figure, above. These microcracks can expand to larger flaws, which lower resistance to erosive and abrasive wear. Using EDM to finish samples can allow for at least rough shaping without the introduction of flaws.

Task 3. Application-specific Testing

AlMgB$_{14}$-based composites outperform commercial WC and cBN materials in ASTM dry erosion and wet abrasion tests. For this reason, AlMgB$_{14}$-based composites are being evaluated for their potential as bulk pumping seals, advanced machine tooling inserts, and abrasive waterjet machining components. Abrasive water jet cutting nozzles are now commonly made of Roctec 100 with sapphire or diamond orifice inserts. Roctec 100 is fine-grained tungsten carbide produced without a binder metal by rapid omnidirectional consolidation. It is substantially more wear-resistant than standard WC-Co cermets.

Missouri University of Science & Technology has tested two wear-resistant boride compositions in their high-pressure waterjet laboratory. A photograph of the test setup is shown in Figure 3.1.1.

Figure 3.1.1. Abrasive waterjet test configuration at Missouri University of Science & Technology. The sample under test is the square-shaped piece near the center of the photograph. Test conditions: conventional abrasive cutting head - sapphire orifice 0.010", mixing nozzle 0.030", abrasive - garnet 80mesh, abrasive feed rate 0.62 lb/min.

The jet consists of garnet particles (nesosilicates) entrained in a high-pressure (40 ksi) stream of water, which impinges on the sample at an angle of 10°. Samples of commercial Borazon 7000 cBN and Roctec 500 WC were tested along with AlMgB$_{14}$-TiB$_2$ samples prepared at the Ames Laboratory. Post-mortem analysis of the mixed-phase boride samples that were tested revealed that failure was due to fracture instead of wear. This indicates that the compacts contain pre-existing flaws or defects, which reduce the material's resistance to shock loading. Examination of one of the compacts (discussed in detail below) revealed the presence of radial stress fractures around the circumference, which can act as nucleation sites for fracture when the 40 ksi abrasive waterjet test is started. These defects may be caused by a combination of end capping and ring capping, dry pressing defects aggravated by the presence of differential springback near the ends of the compact. Springback refers to an increase in one or more dimensions of a pressed part upon ejection, caused by stored elastic energy. The relatively high compaction pressures employed during hot pressing of these ceramic parts (~ 100 – 110

MPa) exacerbates the problem of differential springback. Several factors contribute to this phenomenon, including adhesion of the compact to the punch, excessive spacing between the die wall and punch (i.e., larger than the average particle diameter), and non-uniform shrinkage during compaction due to density gradients. The escape of entrapped gas during evacuation can draw particles into the gap between die and punch, increasing springback near the edges. Die wall lubrication such as grafoil is commonly employed to reduce die wall friction and the associated springback, and is routinely employed for dry pressing of the boride composites. The presence of excessively large particles or granules can also contribute to differential shrinkage and density gradients.

As stated, microscopy of the wear craters after the waterjet testing revealed large sections of material that fractured. As shown in Figure 3.1.2, numerous cracks and fracture surfaces can be seen, many of which do not contain debris from the erosion testing (dark material). This indicates that these fracture surfaces had little interaction with the jet itself and may have been the result of pre-existing flaws. Thus macro-scale failure mechanisms dominated over micro-scale erosive wear.

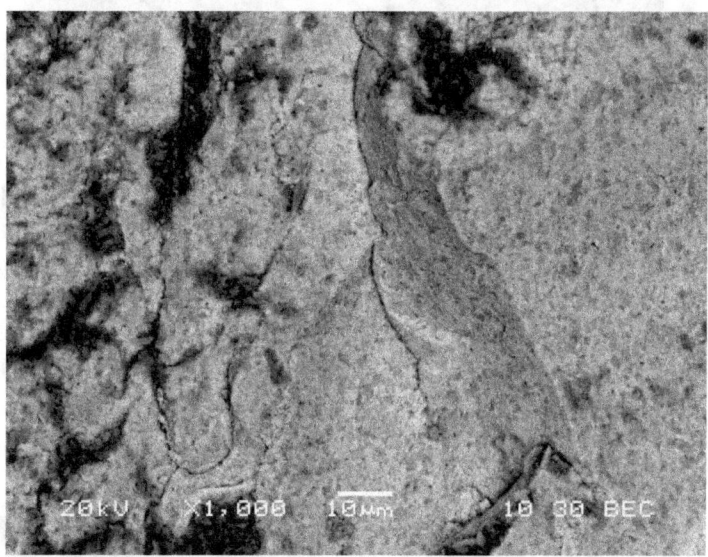

Figure 3.1.2. Abrasive water jet erosion crater mixed-phase boride sample pressed at 1500°C. Cracking and bare fracture surfaces are visible and free of wear debris. Black material is residual debris from testing.

The presence of radial cracks, discussed above, can be seen in the optical micrograph in Figure 3.1.3. The presence of these cracks had not previously posed an issue because the microhardness, dry erosion, and wet abrasion tests were all performed near the center of the samples. Pre-existing flaws of this nature could likely cause or exacerbate the catastrophic damage shown in Figure 3.1.2. This is related to the discussion entitled "Surface Preparation" where similar flaws were introduced by aggressive milling techniques.

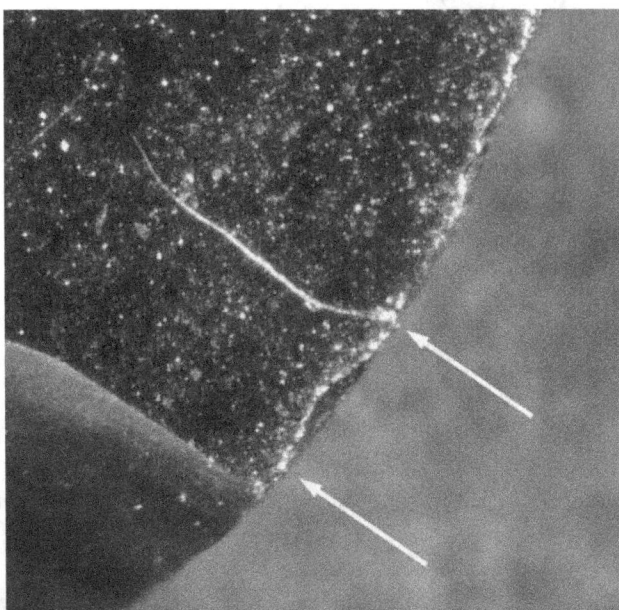

Figure 3.1.3. Pre-existing radial cracks (arrows) around perimeter of a hot pressed AlMgB$_{14}$ sample, one of which resulted in fracture.

The short erosion time (~2 s) was found to produce relatively little damage to either the WC (RocTec) or cBN samples. Wear mechanisms were not apparent in the cBN sample, yet the initiation of failure was seen in the monolithic WC. Figure 3.1.4 shows the pattern of material loss around each particle boundary. Higher magnification (on the right) shows small-scale grain boundary failure within each "supergrain" of finer particles. Grain boundary failure is undesirable because it absorbs less energy and the local nature of the failure may lead to supergrain pull-out/dislodgment at longer erosion times, which would result in increased wear rates after an early run-in period. Similar failure in the form of grain ejection is seen after dry erosion (Figure 3.1.5).

Figure 3.1.4. Abrasive water jet erosion crater in Roctec 500 WC.

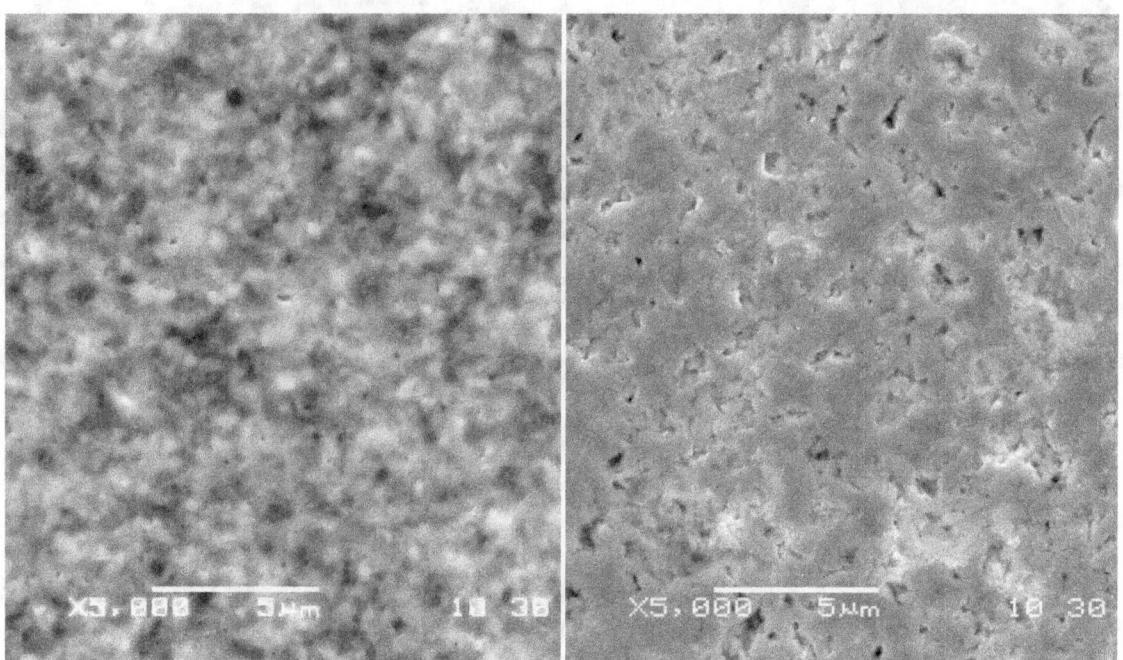

Figure 3.1.5. Left: high contrast micrograph of Roctec 500 WC showing grain size.
Right: dry erosion crater of same sample showing grain ejection.

Waterjet testing had previously been conducted on an archive $AlMgB_{14}$ - 70wt% TiB_2 sample for an extended period of time (68s). Here, the waterjet was centered on the sample, instead of near the edges (which may contain radial cracks). For comparison, micrographs of its wear crater were also taken. Even after the longer testing time, no evidence of large scale fracture was observed. Wear appears to occur due to small scale failure events with thin layers or flakes of material being removed. This failure mechanism is very similar to that observed after dry erosion testing (Figure 3.1.6).

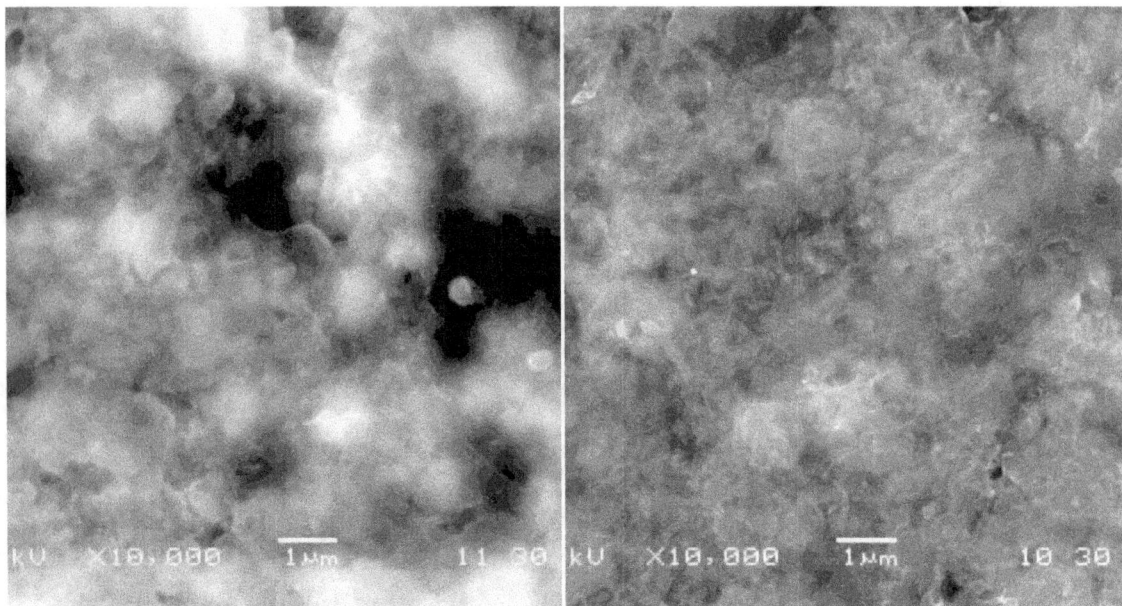

Figure 3.1.6. Left: Abrasive water jet erosion crater of an early series AlMgB$_{14}$ - 70wt% TiB$_2$ sample. Right: Dry erosion crater of similar sample. Note: higher contrast in left image due to bonded erodent material.

The formation of pre-existing flaws in the hot pressed compacts is one of the primary reasons for development of alternate compaction technologies such as Dynaforging and hot isostatic pressing, both of which should eliminate the defects caused by uniaxial dry pressing.

The first batch of samples tested for wet abrasion resistance contained pre-existing flaws from hot pressing, which resulted in fracture during the tests. A second set of samples was prepared by hot isostatic pressing, which should eliminate radial cracks due to unbalanced hoop stresses during consolidation. In August, 2009, we began seeking a manufacturer of abrasive water jet cutting nozzles who would work with us to fabricate and test a AlMgB$_{14}$ - TiB$_2$ nozzle. Three nozzle manufacturers were approached about this possibility, and one of these, Accustream Waterjet Products in New Brighton, MN agreed on September 2 to participate with Ames Lab to fabricate and test a prototype boride nozzle. Unfortunately, an NDA was never agreed upon so the project never started.

Samples of AlMgB$_{14}$/TiB$_2$ were consolidated by hot isostatic pressing (HIPed) in order to avoid the radial cracking problem described above. The samples were held at 1400°C and 300 MPa for 1 hr. This method could potentially generate samples large enough to fashion wet and dry erosion nozzles. The HIPed samples possessed a slightly finer, more angular microstructure than their counterparts uniaxially hot pressed at 1500°C and 100 MPa, yet were otherwise very similar in morphology and density. Density values are listed in the following table.

Table 3.1.1: Density of HIPed TB50 and TZB samples. TB50 is a composite containing 50% TiB_2. TZB is a mixed TiB_2 and ZrB_2 material.

	Temp	Density	
	°C	g/cm3	%
HP-TB50	1500	3.57	~99%
HIP-TB50	1400	3.59	~99%
HP-TZB	1400	5.11	94.8%
HP-TZB	1450	5.34	99.0%
HIP-TZB	1400	5.33	99.0%

Local waterjet cutting facilities were used to test these specimens. These local tests were intended to complement those of our partner, Missouri University of Science & Technology. One key objective was to determine if the recent process improvements had eliminated edge fracturing so that the jet would not immediately expand pre-existing radial cracks in the samples. The waterjet system was set for cutting steel, with parameters listed in Table 3.1.2, as it would likely be aggressive enough to cause measurable damage without the risk of cutting or excessively fracturing the samples. Also included in the table are the typical settings for our ASTM dry erosion tests to illustrate the aggressiveness of the testing. Water can be more easily pressurized than gasses, thus wet erosion is typically a more aggressive environment and is used in cutting applications, while abrasive gas jets (dry erosion) are more often employed for cleaning and surface treatments.

Table 3.1.2: Parameters for Waterjet and Dry erosion conditions.

	Waterjet	Dry Erosion
Abrasive Flow Rate	270 g/min	2-5 g/min
Fluid Pressure	40,000 psi	20 psi
Abrasive Velocity	335 m/s	70 m/s
Jet Speed/Dwell Time	0.6 cm/s	3600 s

By affixing all samples to one expendable steel substrate, the testing of over two dozen samples required only 2 minutes of cutting time. Unlike the laboratory-scale ASTM dry erosion system, the waterjet was not fixed at one point on each sample and instead traversed across each sample surface at a constant rate.   Because of the extreme wear resistance of the boride compacts, the abrasive waterjet did not actually result in a sectioning of the samples, even though the steel holder the samples were mounted on was completely cut through during the tests (where not protected by the samples Rather, passing the jet over the compacts produced a barely visible wear track. By measuring the cross-sectional area of each scar with a profilometer, the volume per unit length of the scar could be calculated. Since the flow rate of the erodent is also known, measurements of volume loss per unit mass of erodent can be calculated, just as in ASTM dry erosion testing. To illustrate the relative effect of impact during high-pressure waterjet erosion, before and after images of the garnet grit are shown in Figure 3.1.7. The damage to alumina particles of similar morphology after dry erosion is not as extensive; before and after images are quite similar.

Figure 3.1.7. Garnet grit used as abrasive in water jet cutting, before (left) and after (right) a single use.

Table 3.1.3 lists the calculated waterjet erosion rates of various samples, along with the corresponding rates measured by ASTM dry erosion tests. The aluminum oxide sample was a high-density (3.69 g/cc; zero porosity) alpha-phase specimen, typically employed for wear-resistance and ballistic armor applications. Figure 3.1.8 compares a profilometer scan across the wear track of one of the mixed-phase boride samples (TB80) with that of the high-density alumina specimen. The figure clearly shows the superiority in erosion resistance of the boride material, indicated by the more shallow profile across the scar.

Table 3.1.3. Wet and Dry erosion rates for select samples. In the Sample column, "AL" = Ames Lab, "TBxx" is $AlMgB_{14}$ + xx% TiB_2, and "NTC"=NewTech Ceramics.)
(Lower values for erosion rate correspond to better performance)

Sample	Sample	Waterjet Erosion	ASTM dry Erosion
Full name		mm^3/kg	mm^3/kg
SAMB-0STB-AW-9	AL TB00	55.4	2.37
SAMB-30STB-AW-8	AL TB30	45.7	2.61
SAMB-STB70-AW-5	AL TB60	31.1	1.36
SAMB-JS4-TB80-L1t	AL TB80	13.4	0.25
RC15-88A-1400-M30	NTC TB50	33.2	0.56
Alumina	Alumina	320	-

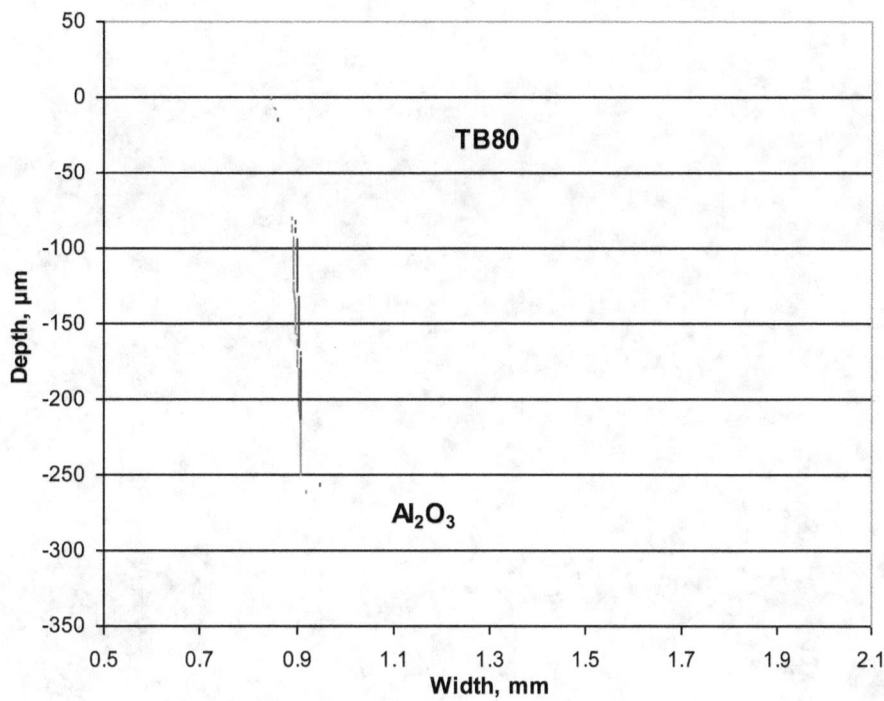

Figure 3.1.8 Comparison of profilometer scans across waterjet erosion track of high-density alumina (red curve) and TB80 boride composite (blue curve). Wear volume is proportional to the depth of the curve.

The NTC sample was a compact hot pressed at Ames from powder provided by NewTech Ceramics, the corporate licensee of the patents protecting the IP of this project's technology. These tests reflect radically different environments, and the erosion rates are seen to reflect the different mechanisms taking place within the materials. While the waterjet erosion rates are comparable between AL-TB60 and NTC-TB50, there is a significant difference in their corresponding dry erosion rates. Differences in processing, composition, and microstructure can lead to observable changes in erosion mechanisms. Microscopy of the waterjet erosion scars reveals differences within the various Ames Lab samples corresponding to different processing (Figure 3.1.9). The largest morphological differences are "gouging" modes in sample TB00 versus "cratering" in sample TB80. Sample TB80 exhibited the best wet erosion resistance of all the boride samples, despite the large craters seen in Figure 3.3.

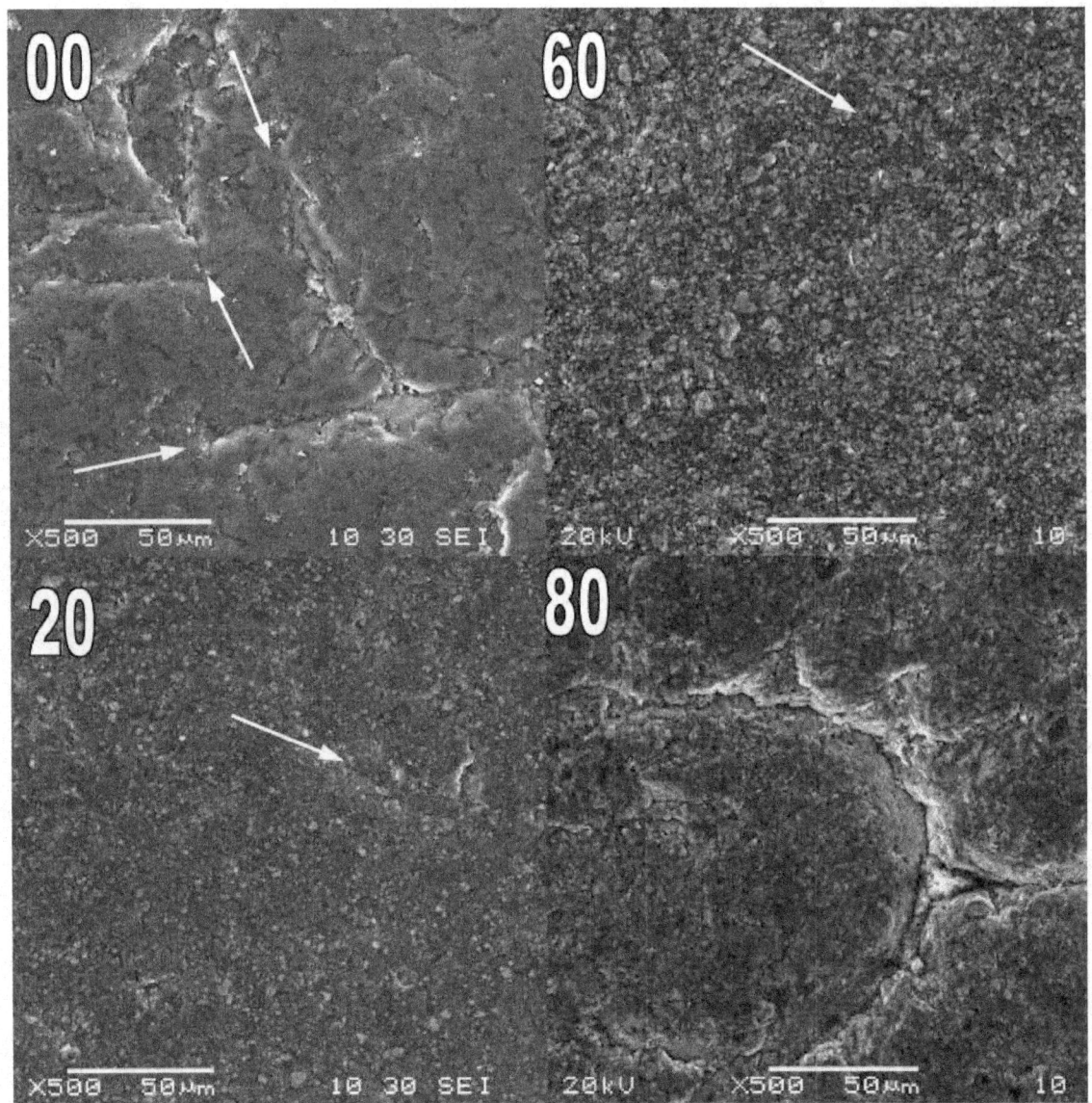

Figure 3.1.9. Waterjet erosion surfaces of TB00, TB20, TB60, and TB80. Samples TB00-60 were milled 30 min with stearic acid additions, sample TB80 was milled 12 hr with no additions. Arrows indicate "gouges" (some faint).

Samples TB00 to TB60 show increasing resistance to "gouging" with increasing TiB_2 content. Gouging appears to result from impact with the edges of highly angular erodent particles. Sample TB80, prepared by a different method (longer milling, higher Fe content) and with higher TiB_2 content, has exhibited higher hardness and lower wet and dry erosion rates. Yet visually the accumulated damage appears to be causing large-scale failures. It is interesting to note the circular fracture pattern in the figure. This failure appears not to be caused by the gouging mode seen in the other samples. In contrast, the high hardness of this sample may be causing the erodent particles to deform until the contact area increases to roughly the size of the feature shown, at which point the sample begins to plastically deform. Like a spherical hardness indentation, there exists stress concentration and deformation at the perimeter of the indent/impact, and this highly deformed material is readily removed by subsequent impacts. Another contributing factor is the associated thermal effects, in combination with steam generation. Despite the rapid cooling from the jet, high local temperatures are achieved (this is visually confirmed from the glowing of the sample from within the water jet). If this is correct, cracking at the edge of the indent-like impact may be infiltrated by water that is heated by subsequent impacts. This would cause microscopic steam explosions and could account for the relatively deep appearance of the cracks in sample TB80. The higher concentrations of Fe in the TB80 sample may make it more susceptible to chemical attack by water, facilitating water penetration into the sample.

It is not yet apparent what microstructural characteristics cause cratering or gouging to dominate. As mentioned, there are differences between waterjet and dry erosion of Ames Lab and New Tech samples. New Tech samples have been found to contain a W-rich third phase, while most of the Ames Lab samples contain an Fe-rich phase. In previous tribology studies, these borides are known to form lubricating layers of boric acid. These reactions may be accelerated as a result of the heat generated by the impacting jet. Each sample was observed to emit visible radiation during contact with the jet, indicating that local temperature excursions on the order of 600 to 700°C were encountered. A possibly useful experiment would be to expose samples to waterjet testing without the presence of the abrasive grit, in order to better understand the nature of the chemical and thermal effects of the high pressure water jet alone.

Recent industrial inquiries concerning new uses of $AlMgB_{14}$ + TiB_2 composites have motivated studies of the corrosion resistance of the composites in a sulfuric acid (H_2SO_4) environment. Corrosion tests in concentrated sulfuric acid were performed on material produced by both Ames Laboratory and NewTech Ceramics (NTC). Table 3.1.4 summarizes the mass loss and corrosion penetration rate for each sample. The TB100 pure TiB_2 was found to corrode much more rapidly than the composite materials. The Ames Laboratory-produced samples were both prepared identically and demonstrate that the composite's performance is superior to that of phase-pure TiB_2. Material prepared by NTC has previously exhibited variations in erosion rates due to differing impurity content introduced during large-scale powder production. These variations in impurities are the likely cause for the difference in corrosion rate among the NTC samples. The composite materials appear to be suitable for corrosive environments containing sulfuric acid. As with other applications explored for these composites, impurity content must be controlled to obtain the highest degree of performance.

Table 3.1.4: Mass loss and calculated corrosion penetration rate for samples treated with concentrated sulfuric acid for 90 hr.

Sample	ΔM, g	Rate, mm/yr
NewTech TB50 (1)	0.0013	0.131
NewTech TB50 (2)	0.0003	0.032
AmesLab TB60	0.0008	0.053
AmesLab TB100	0.0040	0.250

An extension of the dry erosion tests was performed at higher velocity. Composite samples were subjected to a higher pressure gas stream imparting abrasive particle velocities of 110 and 120 m/s particles (compared with 77 m/s used in the earlier tests). The corresponding erosion rates increased to 0.9 mm^3/mg at 120 m/s compared with 0.2 mm^3/mg at 77 m/s. Microstructural examination of the samples was not yet completed at the conclusion of this project. However, it is clear that spinel and residual porosity in the compacts contributes to the erosion rates at the higher velocities.

Task 4. Commercialization

Production-scale manufacturing of this technology was initiated through a start-up company, NewTech Ceramics (NTC). A licensing agreement between the technology owner, Iowa State University Research Foundation, ISURF, and NTC was executed and the company began seeking start-up funds to support initial production and distribution of $AlMgB_{14}$-TiB_2-related material to potential customers. While production levels are proprietary information, NTC has successfully achieved multi-kilogram processing capabilities. The company has opened a production plant in Boone, IA (in close proximity to the Iowa State University campus) This project included providing technical support to NTC in order to facilitate cost-effective mass production of the wear-resistant boride components. A series of technical reports were prepared, detailing the results of our study of reduced-energy processing of $AlMgB_{14}$-related nanocomposites and of characterization of NTC materials. These reports were distributed to NTC to assist in their commercialization efforts. A number of findings were highlighted in these reports. For example, by reducing the milling time from 12 hours to 1 hour and assuming the use of a 1/3 HP motor, the energy requirement for processing is reduced from 3388 Btu/gm to 282 Btu/gm, which translates to an annual energy savings of 3.1×10^9 Btu for a 1,000 Kg production level. The results and implications for large-scale production were discussed with NewTech personnel during occasional meetings held throughout the duration of this project.

NTC has begun working with several major corporations across the country on developing new energy-saving applications for the wear-resistant boride material, based on results obtained from this program.

Task 5: Computational Modeling

5.1: Wear Modeling

A robust computational model is needed to guide experimental efforts and to aid in the fundamental understanding of the wear mechanisms in these complex materials. The modeling effort will initially focus on the role of the ceramic reinforcement phase and the effect of varying volume fraction and size distribution.

The initial work on computational modeling showed the following:

1) The optimum volume fraction of reinforcing phase is dependent on the abrasion condition. When the ratio of the abrasive size to that of the reinforcing phase is decreased, more reinforcing phase must be added to maintain the minimum wear loss.

2) A combination of large and fine reinforcing particles is more effective to resist wear than single-size reinforcing particles. The optimum combination of fine and coarse reinforcing particles is dependent on the abrasion condition. When the ratio of the abrasive size to that of the coarse reinforcing particles is decreased, adding more fine particles is beneficial.

It is generally understood that porosity is detrimental to the wear resistance of materials. However, some experiments show contradictory results or exceptions. A study was conducted to investigate the influence of porosity on the solid-particle erosion of composite materials with the aim of clarifying relevant issues. The modeling study demonstrates that porosity could improve the performance of materials under some conditions. In order to better understand mechanisms involved and generate guidelines to take advantage of porosity, solid-particle erosion processes of composite materials having different pore densities, volume fractions and orientations of reinforcements were studied. The synergic influence of reinforcement/matrix interfacial bond strength and porosity on erosion resistance was also modeled for detailed information. Consistency between the modeling and previous experimental observations was found.

A paper discussing the effect of porosity on wear of composite materials was presented at the Wear of Materials conference, WOM2009, and it attracted considerable interest. Many questions centered around the manner in which porosity would affect the wear behavior of pure metals and ceramic materials, since the answer would help to better understand the general role of porosity in wear processes. Based on the feedback from the audience, we recently simulated the response of a pure metal (Cu) containing pores in comparison with a CrC/Cu composite. It was demonstrated that a small amount of porosity improved the wear resistance of the ductile material, as illustrated in Figure 5.5.1. From the calculated results, one may clearly see that for brittle materials (i.e., ceramics), porosity is always detrimental, but the situation changes for ductile materials.

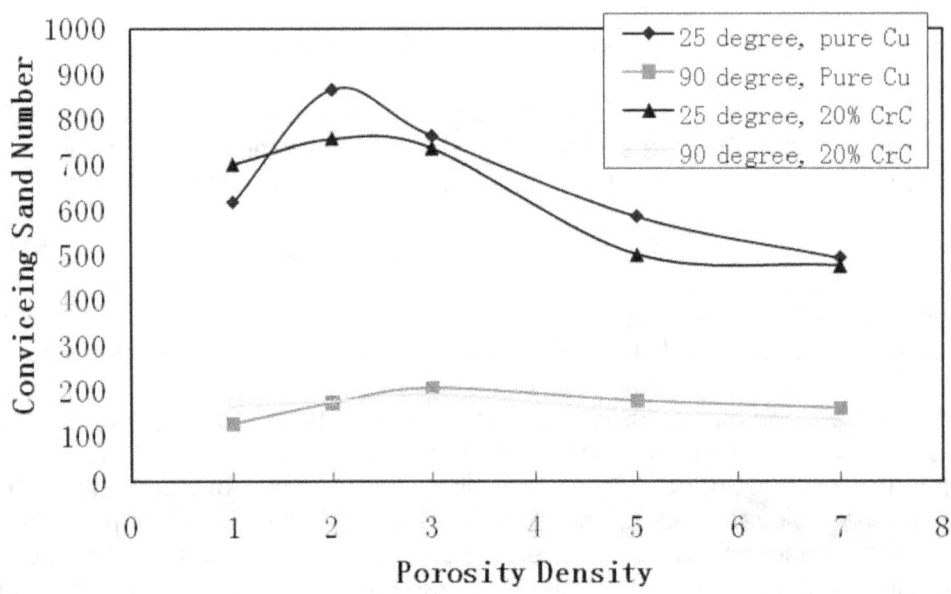

Figure 5.5.1. Effect of porosity (percentage) on the minimum number of erosive impacts necessary to initiate material removal from a pure metal or MMC. A higher value corresponds to greater erosion resistance.

Continued work on the computational modeling generated four additional findings:

1)   Porosity is not always detrimental; a small amount of pores appear to be beneficial to the erosion resistance of the modeled composite. With an increase in porosity, the erosion resistance increases initially and then decreases.
2)   The beneficial effect of porosity does not exist when the volume fraction of reinforcements is high. The erosion resistance continuously decreases as the porosity content increases.
3)   The reinforcement orientation affects the porosity's role and thus the erosion resistance. It was shown that lateral reinforcements had the highest erosion resistance while vertical reinforcements had the lowest erosion resistance.
4)   When the interfacial bond between the reinforcement and matrix is weak, the erosion resistance of the composite is low, and the effect of porosity on the erosion resistance is always negative.

Previously we investigated the effect of porosity on wear of MMCs and demonstrated a positive effect of porosity on the wear resistance of MMCs when the amount of porosity is small. This work was extended to a two-phase ceramic matrix composite (CMC). In the present study, a TiB_2-20%TiC ceramic composite eroded by SiC is simulated using a micro-scale dynamic approach developed by the Surface/Wear group at University of Alberta. The results are expected to be qualitatively similar to those for TiB_2-$AlMgB_{14}$.

Subsequently, the adjustable parameters of the micro-scale dynamic model were modified to account for the known properties of $AlMgB_{14}$ and TiB_2. Some of the required mechanical property values were unknown until measured or calculated. For example, elastic modulus was measured at Ames Laboratory using ultrasonic wave attenuation, and fracture strain was estimated using the Tabor approximation, which states that the Vicker's hardness, Hv, is related to the yield stress, Y through the relationship Hv = 3Y. Table 5.5.1 lists the moduli values and corresponding fracture strain values for the various compositions of interest.

Table 5.5.1. Elastic moduli values obtained on various $AlMgB_{14}/TiB_2$ compositions by ultrasonic measurements at Ames Laboratory and corresponding calculated yield strain.

TiB_2 wt %	Young's (GPa)	Longitudinal (GPa)	Shear (GPa)	Bulk (GPa)	Poisson's ratio ν
0	375	404	160	190	0.17
30	411	440	176	205	0.16
50	440	468	190	215	0.16
70	483	509	212	226	0.14
100	509	527	227	224	0.12

	Hardness (GPa)	Hardness Vicker's	Yield Str (MPa)	E (GPa)	Yeild Strain
0	29	2957	986	376	0.26%
10	30	3059	1020	390	0.26%
20	31	3161	1054	404	0.26%
30	33	3365	1122	418	0.27%
40	34	3467	1156	432	0.27%
50	35	3569	1190	446	0.27%
60	36	3671	1224	460	0.27%
70	38	3875	1292	474	0.27%
80	36	3671	1224	488	0.25%
100	33	3365	1122	516	0.22%

Figure 5.5.2 shows that with an increase in the porosity concentration (vol% pores), the conceiving number (a measure of the number of impacts necessary to generate material removal) decreases regardless of the impingement angle, representing a continuous decrease in resistance to erosive wear. At higher impingement angles, erosion damage is enhanced, but the effects of porosity on wear of the ceramic material at different impingement angles are similar. Different from MMC, ceramic matrix composites are brittle, susceptible to defects, and have low capability to absorb impact energy. The result of the simulation is expected.

Figure 5.5.3 illustrates cross-sections of damaged samples. As shown, there is little plastic deformation. The pores in the ceramic material become the origins of cracks, which propagate extensively along the vertical direction. This is different from the behavior of ductile materials. The adjustable parameters can be modified to account for the unique properties of $AlMgB_{14}$.

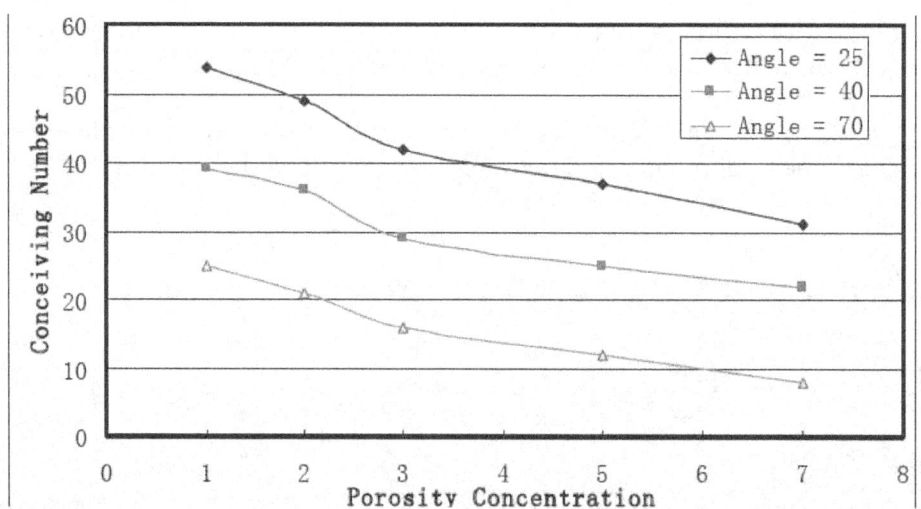

Figure 5.5.2: The conceiving number of a CMC with varying porosity concentration for different impingement angles. A higher conceiving number correlates to higher wear resistance.

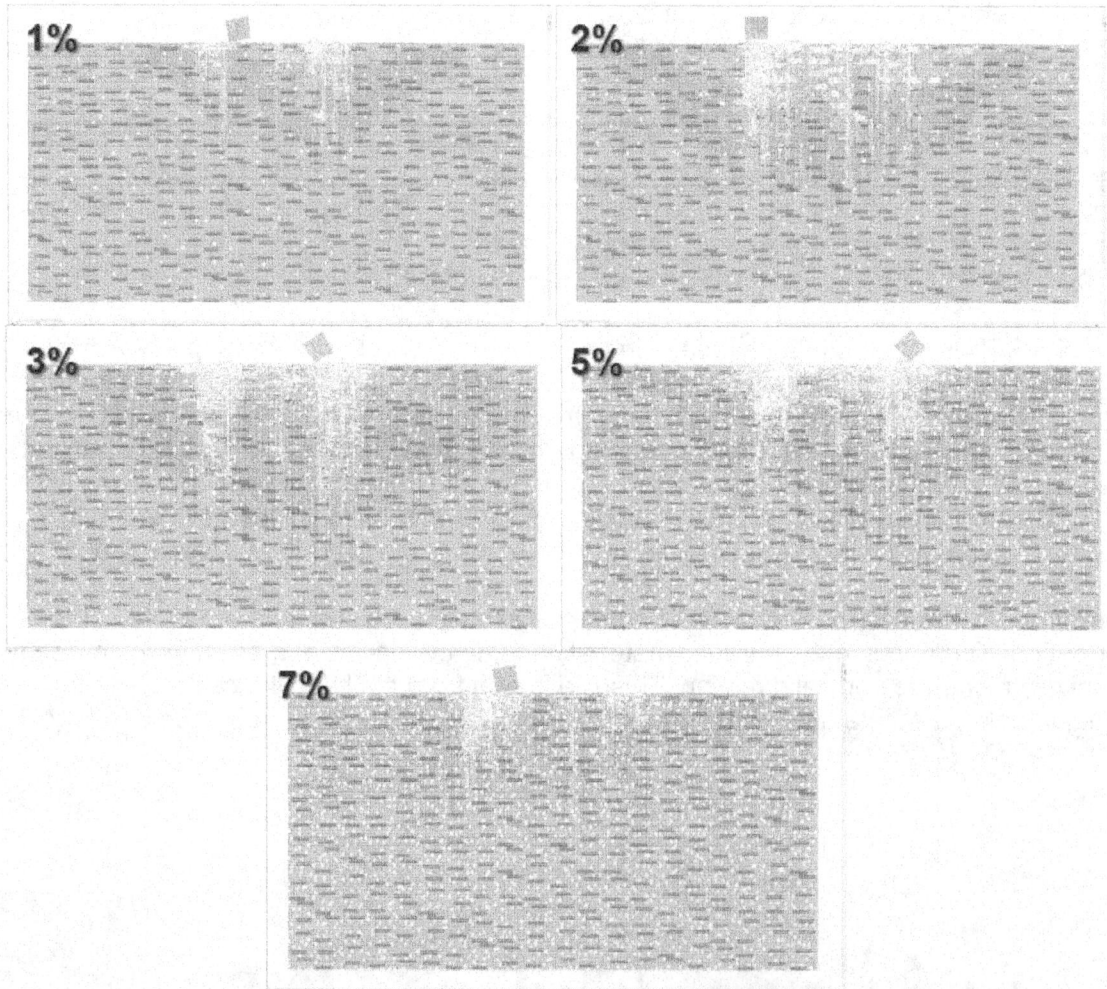

Figure 5.5.3: Simulated erosion damage for different porosity concentrations within a CMC. Impingement angle: 40 degrees.

This study was extended to investigate how the reinforcing phase and porosity affect the performance of a CMC composite ($AlMgB_{14}$ - TiB_2) during solid-particle erosion. $AlMgB_{14}$ and TiB_2 comprised the sample matrix and reinforcing phase, respectively, with silicon carbide was selected as the erodent. The following results were observed:

1) In the ceramic-matrix composite, porosity is <u>always</u> detrimental.
2) The reinforcing phase increases the resistance of the material to solid-particle erosion.
3) A horizontal arrangement of the reinforcements benefits the material the most.
4) Little plastic deformation occurred during the erosion process. Microstructural damage was found to result from cracking rather than from the cutting that usually occurs in ductile materials

5.2: Electronic Structure Modeling

We also investigated late-stage compositional improvements prior to the conclusion of the project. One approach is to exploit recent density functional theory (DFT) results that suggest $AlMgB_{14}$ can be made much harder with substitution of Fe. Sc addition may also enhance stability of the material.   In brief, $AlMgB_{14}$ is quite different from boron carbide and other boron-rich solids. Unlike the C, N, and P atoms in conventional boron-carbide-type structures, the metal atoms in $AlMgB_{14}$ do not share strong covalent bonds with the B-network.[Kolpin,Lowther] Instead, the valence electrons of the Al and Mg atoms become delocalized within the B-network. An ISU MSE faculty member, Prof. Scott Beckman, has begun development of a computational model for the electronic properties of $AlMgB_{14}$. The electronic density of states (DOS) near the band gap is associated with the B-network, so the metal atoms do not influence the band gap. When an Al or Mg atom is removed, the valence electrons that are delocalized within the crystal are removed and the Fermi energy (E_F) shifts within the DOS. The calculated DOS for stoichiometric $AlMgB_{14}$ and off-stoichiometric $Al_{0.75}Mg_{0.75}B_{14}$ is shown in Fig. 5.5.4(a). Removal of the Al and Mg atoms from the 64 atom unit cell does not change shape or energy of the DOS, but instead shifts the E_F from above the conduction band to the valence band, corresponding to the removal of the five valence electrons. It is also found that the crystal can be doped by substitution on the metal atom sites. For example, replacing the four Mg atoms with Li reduces the number of electrons by 4 per supercell, which shifts E_F to the top of the conduction band as shown in Fig. 5.5.4(b).

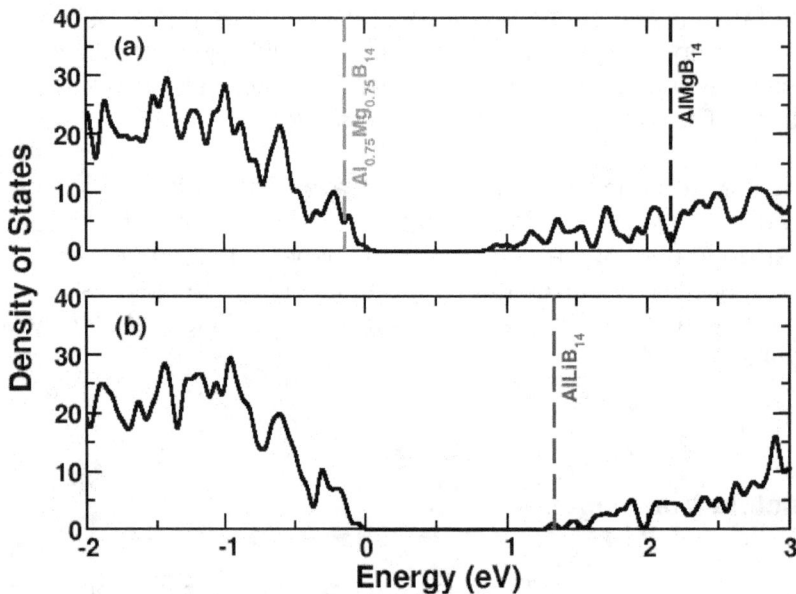

Figure 5.5.4. Calculated density of states and Fermi energy (E_F) of $AlMgB_{14}$ and off-stoichiometric $Al_{0.75}Mg_{0.75}B_{14}$ (1a); Calculated DOS and E_F of $AlLiB_{14}$ (1b).

Based on this approach, the removal of one electron past the $Al_{0.75}Mg_{0.75}B_{14}$ level, to -6 electrons, causes huge changes to the structure and mechanical properties. The crystal's volume shrinks by 11% and the bulk modulus increases by 20% from 200 to 251 GPa. When the Fermi level in Fig. 5.5.4 dips into the valence band, there is a transition from the nominal phase (NP) of $AlMgB_{14}$ to the superhard phase (SP). It is well known that the energetically preferred structure is the off-stoichiometric $Al_{0.75}Mg_{0.75}B_{14}$. It is only a matter of removing one electron (or possibly less) from the off-stoichiometric compound to manifest the superhard properties.

If relatively minor composition and processing modifications can produce a significant improvement in wear resistance without adding significantly to the cost (cost/benefit ratio <1), such effort will ultimately contribute to the commercialization potential and utility of this material in energy-saving applications.

Task 6. High Load Indentation Testing

Recently, a number of research papers have been published on high hardness materials [Chung 07]. These materials have received attention due to the "ultrahard" designation based on measurements taken at low loads. Hard materials frequently contain large numbers of small grains. A high density of fine grains generates more barriers to dislocation movement across the grain boundaries. In order to perform accurate measurements, control of the indent-to-grain size ratio is required. A nanoscale grain size material requires a small indenter with a light load to prevent error from grain boundary defects and crack deformation. Thus, a range of instrumentation is necessary to enable various load ranges in order to control the ratio of indent to grain size, since apparent hardness values vary with changing loads on the indenter. Figure 6.1 shows the transition of indentation hardness versus applied load. This figure is partitioned into three primary areas: area (a.) shows that the apparent hardness increases as the load decreases, and area (c.) shows that as load increases hardness reaches a constant value. Most accurate results are obtained at higher applied loads, because a larger deformation area can be sampled. Unfortunately, higher applied loading tends to fracture brittle materials. Thus, it is best to use the maximum applied loads that a material can withstand to measure hardness without extensive fracture.

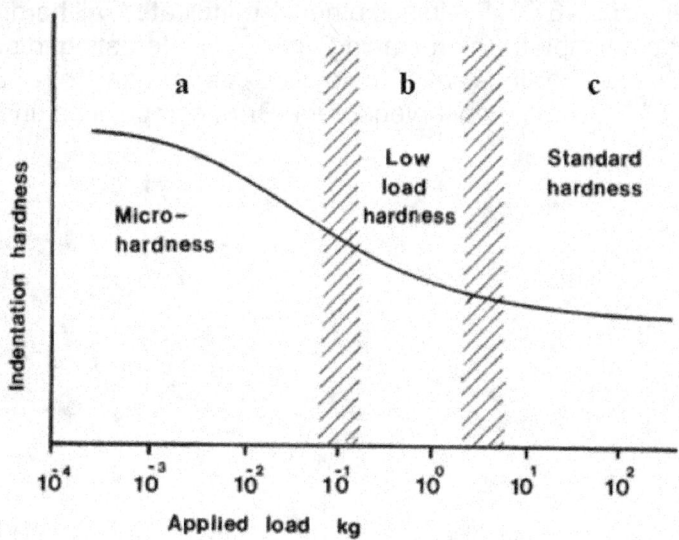

Figure 6.1: Generalized relationship between apparent hardness and applied load.

Figure 6.2 shows a plot of hardness versus load for a SiC standard along with a typical $AlMgB_{14}$ sample. The materials were measured on a Leco-Wilson Tukon Tester with a 500X microscope with optical imaging software. The figure shows the same general trend as depicted in Figure 6.1; as the load decreases the measured hardness values increase. The plots are constructed from average hardness measurements calculated from at least 8 measurements at each load

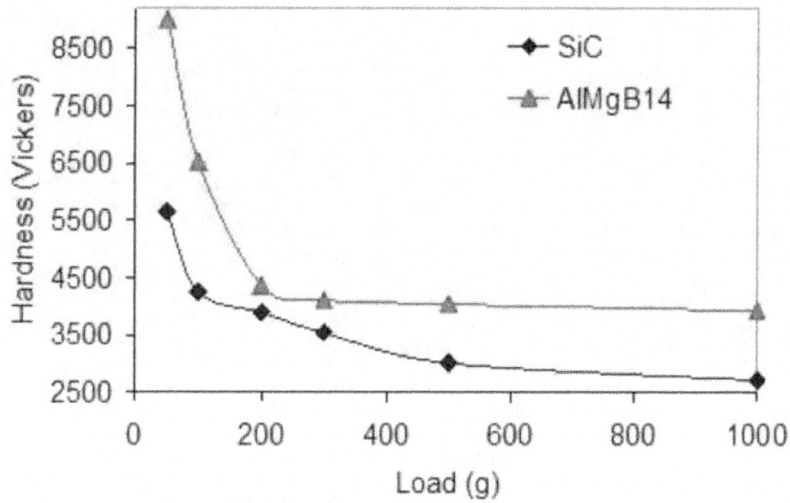

Figure 6.2. Indentation size effect in SiC and a typical $AlMgB_{14}$ compact

The figure shows that the $AlMgB_{14}$ sample possesses a hardness of > 85 GPa (8500 Vickers) under a loading of 100 g (1 N) compared with 55 GPa for SiC. But it is in the asymptotic region of higher loading that the "accepted" hardness value should be defined. Figure 6.3 shows the results of a 5 Kg indentation in an $AlMgB_{14}$-TiB_2 composite. The measured hardness on this sample is 36 GPa, comparable to that of cubic BN under identical loading.   For comparison, the lower micrograph in Figure 6.3 shows a 1 Kg indentation in the same material, with a corresponding hardness of ~ 46 GPa. Unless otherwise indicated, all hardness values in this report correspond to the asymptotic "high-loading" limit. It is interesting to note as a point of comparison that the low-load hardness at 100 g of ReB_2 as reported in [Chung 07] is 40 GPa, which is less than the 64 GPa value observed for the BAM composition under the same loading conditions.

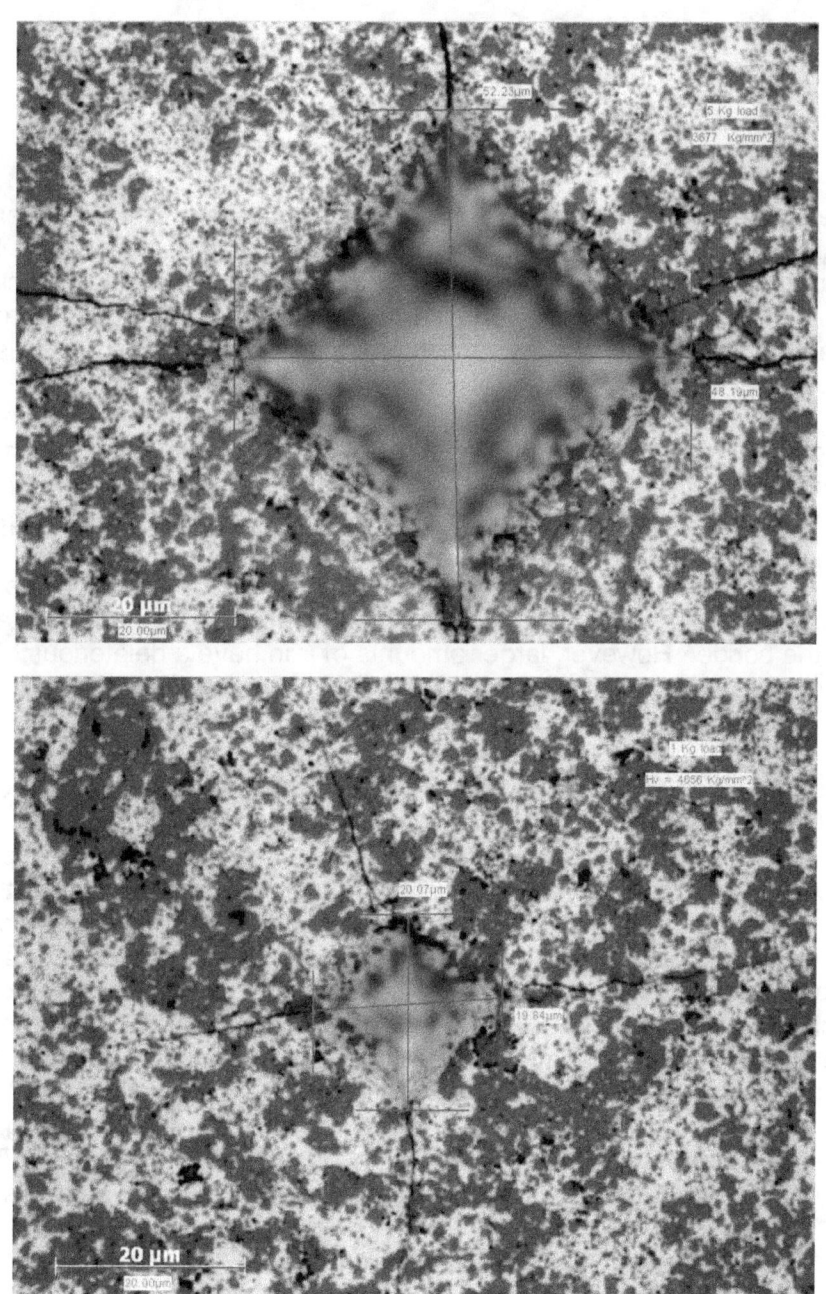

Figure 6.3. Optical micrographs of a 5 Kg indentation (top) and a 1 Kg indentation (bottom) in a typical hot pressed AlMgB$_{14}$-TiB$_2$ composite. The top indent measured at 36 GPa and the bottom indent measured at 46 GPa. The reduction in apparent hardness with decreased loading from 46 GPa at 1 Kg to 36 GPa at 5 Kg is consistent with the indentation size effect.

4.0   Discussion

Over the course of the project, the project participants focused on a deeper understanding of the fundamental material properties of this particular nanocomposite that lead to improved product efficiencies and general performance. Part of the project deliverables included the development and characterization of $AlMgB_{14}$ nanocomposites, establishing reproducible processing parameters, and performance testing of various end applications to establish a measurable energy savings.

Analysis of $AlMgB_{14}$-TiB_2 composites indicate that $MgAl_2O_4$ spinel and iron boride are the major impurities. Neutron activation and inductively-coupled plasma mass spectrometry revealed up to 5 volume percent oxygen and 10 volume percent iron phases. Iron is introduced during mechanical alloying from wear debris of the hardened steel media and vial liner. The major source of oxygen is believed to be surface contamination of the precursor powders. A small amount of iron contamination is actually beneficial because it acts as a sintering aid, increasing densification of the boride. However, larger amounts of iron have a deleterious effect on hardness and wear resistance. The FeB and $MgAl_2O_4$ phases are undesirable because of their lower hardness and fracture toughness vis-à-vis the $AlMgB_{14}$ and TiB_2 phases.

During erosion of WC, the ductile Co binder is compromised by plastic deformation, resulting in unsupported WC grains that fracture and lead to grain ejection. The intersection of fracture surfaces during erosion leads to rapid material removal and grain pullout. Identical conditions imposed on an $AlMgB_{14}$-TiB_2 composite reveals evidence of incipient fracture in the TiB_2 phase. The erosion of brittle materials typically proceeds by repeated fracture from particle impacts followed by the ejection of chips that form as the microcracks coalesce. In the case of the $AlMgB_{14}$-TiB_2 composite, extensive, well-developed cracks are not observed. Examination of samples subjected to higher velocity erosion revealed that the incipient cracks propagate to the TiB_2 grain boundaries and subsequently terminate. This observation suggests that the primary phase responsible for damage initiation is TiB_2. Failure does not occur at the $AlMgB_{14}$-TiB_2 grain boundary, despite the damage within the large grains, again demonstrating the strong bonding between the phases. The fracture and grain pullout of TiB_2 suggests refinement of the TiB_2 reinforcement phase as a viable path to improved toughness and wear resistance.

Another variable that strongly affects erosive and abrasive erosion rates is the processing of the TiB_2. A strong interface between boride phases acts as a reinforcement, increasing both hardness and toughness. For grain sizes on the order of 1 μm or larger, the response of the composite approaches a rule-of-mixtures behavior, whereas a positive deviation is observed in the case of the more highly refined microstructures. Introduction of indentation cracks on and near interfaces between $AlMgB_{14}$ and TiB_2 phases has revealed no indication of delamination or a tendency for the cracks to revert to an intergranular mode from transgranular propagation. The bond strength between $AlMgB_{14}$ and TiB_2 may be enhanced for certain favorable orientation relationships. For example, molecular orbital calculations of the interfacial energy between Al and SiC for various orientations have shown that their bonding strength can exceed the adhesive strength of pure Al. The evidence suggests that the interface between $AlMgB_{14}$ and TiB_2 may provide a critical mechanism for enhanced resistance to wear.

5.0   Accomplishments

This project demonstrated the utility of $AlMgB_{14}$-TiB_2 nanocomposites for a number of wear-intensive applications. Advanced processing methods were developed to improve the performance of the material while reducing the energy intensity for processing large quantities. Results were successfully transferred to the commercialization entity, New Tech Ceramics. Analytical characterization of laboratory scale and commercial powders and bulk articles helped to determine the effects of secondary contamination (metallic and oxide), grain size, and bulk density.   In addition, our analysis revealed the primary source of boric acid in the nanocomposites is the TiB_2 phase and not the $AlMgB_{14}$, providing important guidance for future optimization of wear resistance and lubricity through tuning of the relative amounts of each phase.

In addition, the following patents and publications resulted from this project:

U. S. Patent 7,517,375, "Wear-Resistant Boride Composites with High Percentage of Reinforcement Phase," B. A. Cook, A. Ahmed, A. M. Russell, J. Peters, and J. Harringa, issued April 14, 2009.

U. S. Patent 7,172,641, "Ultra-hard boride-based metal matrix reinforcement," B. A. Cook, A. M. Russell, J. L. Harringa, S. B. Biner, and I. E. Anderson, issued February 6, 2007.

B. A. Cook, J. L. Harringa, J. S. Peters, and A. M. Russell, "Enhanced Wear Resistance in $AlMgB_{14}$-TiB_2 Composites," accepted for publication in the Proceedings of the 2011 Wear-of-Materials Conference (April, 2011 in Philadelphia, PA), and the associated special issue of *Wear*.

B. A. Cook, A. M. Russell, J. S. Peters, and J. L. Harringa, "Estimation of Surface Energy in Solid AlMgB14," J. Phys. Chem. Sol., 71 (2010) 824 .

B. A. Cook, J. L. Harringa, J. S. Peters, and A. M. Russell, "Microstructure and wear resistance of low-temperature hot pressed TiB_2," Wear 266 (2009) 1171-1177.

J. S. Peters, B. A. Cook, J. L. Harringa, and A. M. Russell, "Erosion resistance of TiB_2-ZrB_2 composites," Wear, 267 (2009) 136–143.

Q. Chen, D. Y. Li, and B. A. Cook, "Is Porosity Always Detrimental to the Wear Resistance of Materials? – A computational study on the effect of porosity on erosive wear of TiC/Cu composites," Wear 267 (2009) 1153–1159.

D. Y. Li, Q. Chen, and B. A. Cook, "A further simulation study on the dual role of porosity in solid-particle erosion of materials," Submitted to Wear of Materials Conference 2011 (April 3 – 7, 2011) Philadelphia, PA

D.V.S. Muthu, B. Chen, B.A. Cook, and M.B. Kruger, "Effects of Sample Preparation on the Mechanical Properties of $AlMgB_{14}$," High Pressure Research, 28 (2008) 63.

6.0   Conclusions

Given the scope of the project and the extent to which the AlMgB$_{14}$-TiB$_2$ nanocomposites were investigated, there are several key conclusions that can be drawn. The functional performance testing on production-intent prototype components demonstrates that there are significant gains in the energy/fuel/time/processing efficiency results associated with application of the composite.

The primary focus of this project was the development of AlMgB$_{14}$-TiB$_2$ composites in bulk form for reduced wear and increased energy efficiency in rotating pump interfaces and gas seals, for nozzles in abrasive jet machining systems, and in abrasive slurry transport systems. The foremost goal of material development was the refinement of both composition and microstructure of AlMgB$_{14}$-TiB$_2$ nanocomposites to improve strength and wear resistance. Powder processing and methods of hot consolidation were studied to produce techniques leading to commercially viable manufacturing methods. Key conclusions for the project are:

- Elimination of residual porosity and deleterious contamination phases produced the exceptional hardness/toughness combination of 30-35 GPa and 10 MPa$\sqrt{m}$.
- Intragranular fracture is the primary failure mode for AlMgB$_{14}$-TiB$_2$ composites.
- Minimizing grain size reduces the volume for intragranular fracture which allows crack energy to be dissipated in the interfacial boundaries between phases.
- Wear resistance is enhanced by reducing the occurrence of intragranular fracture through grain size reduction to the nanometer scale.
- Alternate powder processing techniques to mechanical alloying and hot pressing can be utilized to produce bulk composite materials.

The diverse and extensive potential applications for these nanocomposites are expected to offer a unique opportunity to realize significant energy savings as a result of reduced frictional losses and elimination of inefficient operation resulting from the wear of components.   Research involved powder processing, consolidation, and characterization of boron-rich composites, with the objective of achieving a reproducible 30% increase in erosive and abrasive wear over current state-of-the-art wear resistant material such as binderless WC. By eliminating residual porosity and deleterious contamination phases, an exceptional hardness/toughness combination of 30-35 GPa and 10 MPa$\sqrt{m}$ at the conclusion of the project was achieved.

Experimental and computational studies on the effects of second phase additions and ductile binder phase on the erosive and abrasive wear resistance of consolidated articles were performed. Advanced slurry processing, combined with in-situ, self-comminution was developed for laboratory-scale nano-particle synthesis. Oak Ridge National Laboratory provided expertise on novel, large-scale powder production techniques such as microwave and IR heating, on novel consolidation techniques such as spark plasma sintering, and also on tribology testing of AlMgB$_{14}$-TiB$_2$ coatings. Conventional consolidation routes such as pressing and sintering, hot pressing, and isostatic pressing at Ames were also augmented by rapid compaction technology at Carpenter Powder Products, Inc. Field testing of bulk, wear resistant nanocomposites was performed with various industrial partners

From a scientific standpoint, this project determined that a large grained microstructure is more susceptible to intragranular fracture, with much of the crack energy dissipated within the grain.

As the grain size becomes progressively smaller, there exists less volume for intragranular fracture, and more of the crack energy will be dissipated in the interfacial boundaries between phases. But stronger interfacial energies make it more likely that the crack propagation energy can be dissipated without breaking the bonds, thereby preserving the integrity of the interface. If the single phase regions in these composites constitute the weakest link with respect to erosion resistance, then increasing the volume fraction of boride interfaces would contribute to an enhancement in wear resistance. In other words, wear resistance in these unique composites can be enhanced by reducing the occurrence of intragranular fracture through a reduction in grain size to the nanometer scale, thus constraining cracks within the grains and preventing or impeding extensive intragranular propagation, assuming the deleterious effects of oxygen can be controlled.

The key results and lessons learned for this project are summarized as follows:

Primary Results
- Laboratory scale powder production methods can be extended to commercial scale production.
- Bulk materials can be formed by a multitude of readily available consolidation techniques (hot press, hot isostatic press, cold isostatic press/sinter, spark plasma sintering).
- Rapid consolidation techniques such as Dynaforging require higher temperatures to achieve full density.
- A variety of ball mills can be utilized for powder production including commercially available vibratory, planetary, and attritor mills.
- The reduction of phase size and the uniform distribution of phases within the $AlMgB_{14}$-TiB_2 composite are essential to the hardness and wear resistance of bulk articles.
- Minimization of oxygen contamination is important for the production of $AlMgB_{14}$ based materials.
- Electrical conductivity increases with the amount of TiB_2 added, allowing high TiB_2 composites to be machined by electrodischarge machining.

Lessons Learned
- The quality of the precursor materials is important to the quality of the final bulk compact. In particular, boron powder must be treated to remove oxygen and moisture prior to processing.
- Methanol and ethanol are incompatible slurry agents for powders containing unreacted magnesium.
- Small amounts of Fe wear debris (i.e. ~3%) aid in the sintering of $AlMgB_{14}$ based materials.
- Hardness measurements alone are not an indicator for material performance in high wear applications.
- Oxygen exposure of the unreacted powder must be kept to a minimum to eliminate spinel formation.
- As is the case with many powder materials, safe handling of the unreacted powders must be adhered to in order to avoid powder combustion when exposed to air.
- Bulk articles can be commercially produced utilizing standard powder processing and consolidation equipment.

7.0   Recommendations

The following recommendations are offered for further development and implementation of this technology:

1. Depending upon the particular application, the nanocomposite composition can be optimized to enhance either wear or low-friction performance. A composite possessing a higher percentage of the TiB_2 phase would trend towards lower friction performance due to this phase's tendency to form boron oxide, and, subsequently, boric acid. Conversely, in an application driven by the need for increased wear resistance, the percentage of TiB_2 phase could be lowered, as the research has shown that the $AlMgB_{14}$ constituent does not form boric acid.

2. Applications for $AlMgB_{14}$-TiB_2 nanocomposites should implement various test methods as a means to evaluate the materials for performance prior to end-application development. Many process parameters have been shown to influence performance, and so careful process monitoring and predictive evaluation techniques are critical to success.

3. Further enhancements to the $AlMgB_{14}$-TiB_2 nanocomposites could be realized by optimizing the electronic density of states. Initial research has shown that, for example, hardness of the bulk material could be enhanced through stoichometric changes to the overall composition of the ternary compound.

8.0   References

Task 1: Processing & Characterization References

[Atiq 06] A. Ahmed, S. Bahadur, B.A. Cook, J. Peters, Tribology International 39 (2006) 129-137.

[Atiq 09] Ahmed, A.; Bahadur, S.; Russell, A. M.; Cook, B. A.. Tribology International (2009), 42(5), 706-713.

[Peters 09] J.S. Peters, B.A. Cook, J.L. Harringa, A.M. Russell, Wear 266 (2009) 1171-1177.

[Peters 07] J.S. Peters, Improving hardness and toughness of boride composites based on $AlMgB_{14}$. Ph.D. Thesis, Iowa State University, 2007.

[Cook 10] B.A. Cook, A.M. Russell, J.S. Peters, J.L. Harringa, J. of Phys. & Chem. of Solids 71 (2010) 824-826.

[Lewis 01] T.L. Lewis, A study of selected properties and applications of $AlMgB_{14}$ and related composites: Ultra-hard materials. M.S. Thesis, Iowa State University, 2001.

[Cook 00] B.A. Cook, J. L. Harringa, T. L. Lewis, A. M. Russell, Scripta Materialia 42 (2000) 597-602.

[Wang 03] W.M. Wang, H. Wang, Z.Y. Fu, Key Engr. Mat. 249 (2003) 109-114.

[Zdaniewski 87] W.A. Zdaniewski, J. Amer. Cer. Soc. 70 (1987) 793-7.

[Moriyama 98] M. Moriyama, H. Aoki, Y. Kobayashi, Journal of the Ceramic Society of Japan 106 (1998) 1196-1200.

[Millet 96] P. Millet, T. Hwang, Journal of Materials Science 31 (1996) 351-355.

[Peters WOM09] J.S. Peters, B.A. Cook, J.L. Harringa, A.M. Russell, Wear 267 (2009) 136-143.

[Mroz 93] C. Mroz, Amer. Cer. Soc. Bull., 72 (1993) 126.

[Gasch 04] M. Gasch, et. al. Ultra-High Temperature Ceramics, 39 (2004) 5925.

[Weng 09] L. Weng, X. Zhang, W. Han, J. Han, Int. J. Refractory Metals & Hard Materials, 27 (2009) 711.

[Becher 90] P. F. Becher, Annu. Rev. Mater. Sci., 20 (1990) 179.

[Deng 98] J. Deng, X. Ai, Materials Research Bulletin, 33 (1998) 575.

[Deng 05] J. Deng, L. Liu, J. Liu, J. Zhao, X. Yang, Machine Tools & Manufacture, 45 (2005) 1393.

[Kamiya 95] A. Kamiya, K. Nakano, Key Engineering Materials, 108-110 (1995) 219-225

[ORNL] http://www.ms.ornl.gov/researchgroups/process/cpg/sic.htm

[Holleck 87] H. Holleck, H. Leiste, W. Schneider, International Journal of Refractory & Hard Metals 6 (1987) 149-154.

[Torizuka 95] S. Torizuka, K. Sato, H. Nishio, T. Kishi, Journal of the American Ceramic Society 78 (1995) 1606-1610.

[Brazhkin 04] V. Brazhkin, N. Dubrovinskaia, M. Nicol, N. Novikov, R. Riedel, V. Solozhenko, Y. Zhao, Nature Materials 3 (2004) 576-577.

[Chung 07]   H.Y. Chung, M.B. Weinberger, J.B. Levine, A. Kavner, J.M. Yang, S.H. Tolbert, R.B. Kaner, Science 316 (2007) 436-439.

[Kolpin 08] H. Kolpin, D. Music, G. Henkelman, J.M. Schneider, Phys. Rev. B 78 (2008) 054122.

[Levchenko 06] G. Levchenko, et. al., J Solid State Chem. 179 (2006) 2949.

[Lowther 02]   J.E. Lowther, Physica B 322 (2002) 173-178.

[Harmon 02]   Y. Lee, B.N. Harmon, Journal of Alloys and Compounds 338 (2002) 242-247.

[Vajeeston 01] P. Vajeeston, et. al., Phys. Rev. B 63 (2001) 045115.

[Baik 87] S. Baik, P.F. Becher, J. American Ceramic Society, 70 (1987) 527.

[Finch 86] C.B. Finch, P.F. Becher, P. Angelini, S. Baik, C.E. Bamberger, J. Brynestad, Advanced Ceramic Materials, 1 (1986) 50.

[Smallman 99] R. E. Smallman and R. J. Bishop, ""Modern physical metallurgy and materials engineering," 6th ed., Elsevier Science (1999) p. 329.

[MacKenzie 96] K.J.D. MacKenzie, R.H. Meinhold, J. Mater. Chem., 6 (1996) 821.

Task 2: Development of Scale-up Technology References

[Ott 08] R.T. Ott, X.Y. Yang, D.E. Guyer, S. Chauhan, D.J. Sordelet J. Mat. Res 23(1), 2008, 133-139.

[Peters 02] J.S. Peters, B.A. Cook, J.L. Harringa, A.M. Russell Wear 266 (2009) 1171–1177.

Task 6: High Load Indentation Testing References

[Chung 07] H-Y Chung, M. B. Weinberger, J. B. Levine, A. Kavner, J-M. Yang, S. H. Tolbert, and R. B. Kaner, Science 316 (2007) p. 436

www.ingramcontent.com/pod-product-compliance
Lightning Source LLC
Chambersburg PA
CBHW081456170526
45166CB00008B/2451